I0488573

NUREG-1505, Rev. 1

A Nonparametric Statistical Methodology for the Design and Analysis of Final Status Decommissioning Surveys

Interim Draft Report for Use and Comment

Manuscript Completed: June 1998
Date Published: June 1998

Prepared by
C.V. Gogolak*, G.E. Powers, A.M.Huffert

Division of Regulatory Applications
Office of Nuclear Regulatory Research
U.S. Nuclear Regulatory Commission
Washington, DC 20555-0001

*U.S. Department of Energy, Environmental Measurements Laboratory
201 Varick Street, 5th Floor
New York, NY 10014

ABSTRACT

This report describes a nonparametric statistical methodology for the design and analysis of final status decommissioning surveys in support of the final rulemaking on Radiological Criteria for License Termination published by the Nuclear Regulatory Commission in the *Federal Register* on July 21, 1997. The techniques described are expected to be applicable to a broad range of circumstances, but do not preclude the use of alternative methods as particular situations may warrant. Nonparametric statistical methods for testing compliance with decommissioning criteria are provided both for the case in which the radionuclides of concern occur in background and also for the case in which they do not occur in background. The tests described are the Sign test, the Wilcoxon Rank Sum test, and a Quantile test. These tests are performed in conjunction with an Elevated Measurement Comparison to provide confidence that the radiological criteria specified for license termination are met. The Data Quality Objectives process is used for the planning of final site surveys. This includes methods for determining the number of samples needed to obtain statistically valid comparisons with decommissioning criteria and the methods for conducting the statistical tests with the resulting sample data.

CONTENTS

CONTENTS

CONTENTS

FIGURES

FIGURES (cont'd.)

TABLES

TABLES (cont'd.)

TABLES (cont'd.)

FOREWORD

The NRC has amended its regulations to establish residual radioactivity criteria for decommissioning of licensed nuclear facilities. As part of this initiative, the NRC staff has evaluated the application of nonparametric statistical methods as an alternative to the parametric statistical approach described in the U.S. Nuclear Regulatory Commission (NRC) draft report NUREG/CR-5849, entitled, "Manual for Conducting Radiological Surveys in Support of License Termination." The nonparametric statistical approach described in this report is expected to be simpler and more cost-effective for the design and analysis of final status decommissioning surveys when radiological criteria for decommissioning approach background radiation levels. This report also shows the advantages of using the Data Quality Objectives process as it relates to the planning and analysis of final site surveys. The application of the proposed DQO process includes methods for determining the number of samples needed to obtain statistically valid comparisons with decommissioning criteria and the methods for conducting the statistical tests with the resulting sample data.

The initial draft of this report was published in August 1995. As a result of the comments received, extensive revisions were made to include alternative scenarios for statistical hypothesis testing. A number of new concepts have been introduced, and examples for some special cases have been added. The results, approaches and methods described herein are provided for information only and should not be considered a substitute for NRC requirements.

John W. Craig, Director
Division of Regulatory Applications
Office of Nuclear Regulatory Research

ABBREVIATIONS

ALARA	as low as is reasonably achievable
CFR	Code of Federal Regulations
DCGL	Derived Concentration Guideline Level
DOE	U.S. Department of Energy
DQA	data quality assessment
DQO	data quality objective
EMC	elevated measurement comparison
EPA	U.S. Environmental Protection Agency
LBGR	Lower Boundary of the Gray Region
MCA	multichannel analyzer
MDC	minimum detectable concentration
NIST	National Institute for Standards and Technology
NRC	U.S. Nuclear Regulatory Commission
PC	personal computer
PDL	predicted dose level
PIC	pressurized ionization chamber
QA	quality assurance
QC	quality control
TEDE	total effective dose equivalent
WRS	Wilcoxon Rank Sum Test
WSR	Wilcoxon Signed Ranks Test

1 INTRODUCTION

1.1 Overview of NRC Site Decommissioning

At sites and facilities licensed by the Nuclear Regulatory Commission (NRC), the formal decommissioning process begins when a licensee decides to terminate licensed activities. The majority of licenses terminated each year by NRC involve little or no site remediation and, therefore, present no complex decommissioning problems owing to residual radioactivity. However, license termination at a small number of sites is far more complex because contamination may be spread into various areas within the facility and surrounding areas by the movement of materials and equipment, by activation, and by the dispersion of air, water, or other fluids through or along piping, equipment, walls, floors, and drains. Decontamination of such areas is likely to be performed at nuclear power plants, non-power (research and test) reactors, fuel fabrication plants, uranium hexafluoride production plants, and independent spent fuel storage installations. A small number of universities, medical institutions, radioactive source manufacturers, and companies that use radioisotopes for industrial purposes may also contain radioactive contamination that requires remediation.

NRC regulations in 10 CFR 30.36, 40.42, 50.82, 70.38, and 72.54 require licensees to remove their facilities from service safely. As part of the decommissioning process, licensees are required to demonstrate that residual radioactivity in facilities and environmental media has been reduced to acceptable levels. Typically, licensees demonstrate compliance with radiological criteria for license termination by conducting final status surveys of the site or facility and reporting the survey results to NRC for evaluation. Where appropriate, the NRC staff conducts confirmatory surveys to verify that lands and structures have been adequately remediated.

On July 21, 1997, the NRC amended the regulations in 10 CFR Part 20 to include explicit radiological criteria for decommissioning (62 FR 139, pp. 39057–39092). Subpart E of the amended regulations contains dose-based radiological criteria for restricted and unrestricted release, consisting of a total effective dose equivalent (TEDE) limit for residual radioactivity above background. These regulations replace prior NRC guidance based on surface and volume activity concentration limits for specific radionuclides.

To implement the dose criteria in the amended 10 CFR Part 20, final status surveys and confirmatory surveys must be capable of detecting very low levels of residual radioactivity in the presence of background at a variety of NRC-licensed facilities and sites. An essential component of such surveys is a statistical methodology that is appropriate for radiological data at or near background levels. This document presents such a methodology.

1.2 Need for This Report

Previously, the NRC staff used guidance for conducting final status radiological surveys that is contained in draft report NUREG/CR-5849 (1992), entitled "Manual for Conducting Radiological Surveys in Support of License Termination." This report contains an alternative statistical approach for designing radiological surveys. The framework for the survey design is the Data Quality Objectives (DQO) process. The DQO process uses statistical hypothesis testing rather than the construction of confidence intervals. This allows a balance to be reached between the risk of possibly releasing an incompletely remediated site and the risk of possibly requiring further remediation at an already adequately remediated site. One of the primary goals of the DQO process is the determination of acceptable decision error rates for the hypothesis test, i.e., those that will reflect the relative importance of these risks at a specific site. The DQO process is used to incorporate site-specific information and sound scientific judgment into the survey design and data analysis so that the objective of safely releasing a site can be met while reducing the number of unnecessarily arbitrary and conservative assumptions that are sometimes invoked in the face of uncertainty.

Using the DQO framework, the amount and type of data to be collected are related to the specific decision to be made rather than sampling at a fixed density. The number of samples of measurements needed in a survey unit is determined by the acceptable decision error rates, the magnitude of the release criterion relative to the overall variability of the data, and the sensitivity of the scanning method used. The type and amount of scanning required depend primarily on the classification of the survey unit. Three classes of survey units are used to direct the survey effort at a level commensurate with the potential for residual radioactivity in excess of the release criterion. Acceptable areas of elevated activity are determined by radionuclide-specific area factors derived from an appropriate dose model.

The nonparametric statistical techniques described in this report do not require the data to be normally or log-normally distributed and are, therefore, expected to be more appropriate for determining the number of samples required for radiological surveys and analyzing data collected at or near background levels. These tests perform almost as well as the parametric tests even when the data are normally distributed, are less sensitive to outliers, and are better able to handle data sets that include *non-detects*.

There are two possible approaches to demonstrating compliance with criteria that specify a dose limit due to residual radioactivity distinguishable from background, depending on which of the following questions is emphasized:

(1) Does the dose due to residual radioactivity exceed the limit?
(2) Is the residual radioactivity indistinguishable from background?

In the initial draft of NUREG-1505 (August 1995) the approach emphasized question 2. This final report addresses both approaches.

1.3 Objective of This Report

This report describes a nonparametric statistical methodology that NRC licensees may consider when evaluating methods for demonstrating compliance with the radiological criteria for license termination in Subpart E of 10 CFR Part 20. The DQO process (EPA QA/G-4 and QA/G-9, 1994) is used as the framework for the planning of final site surveys. The statistical approach described in this report is expected to be a resource-efficient solution for the design of final status decommissioning surveys when radiological criteria for decommissioning approach background levels. The proposed process includes methods for determining the number of samples needed to obtain statistically valid comparisons with decommissioning criteria and the methods for conducting the statistical tests with the resulting sample data.

No single statistical formulation can adequately anticipate every contingency that will arise in deciding whether a survey unit can be safely released. The DQO process should be used to determine whether a proposed action will further the objective of safely releasing the site. The decisions reached may not always be accompanied by a numerical procedure leading to that decision. However, such decisions should always be accompanied by a description of which actions were taken, and why. The DQO process provides a methodology for resolving the often complex issues surrounding site remediation and decommissioning.

1.4 Structure of This Report

This report is divided into four major parts. The first part deals with general final status survey design criteria, definitions, and data quality objectives (Chapters 2 and 3). The second part describes preliminary data analysis and data quality assessment (Chapter 4). The third part describes the use of the statistical tests recommended in this report (Chapters 5 through 8). These first eight chapters contain all of the information required to design and conduct final status surveys, and to analyze and interpret the results. Chapters 9 through 14 deal with extensions of, and alternatives to, the statistical procedures that may be applicable in some situations. Chapters 15 and 16 contain a Glossary and Bibliography, respectively. The appendix contains the statistical tables needed to perform the analyses described in this report.

2 OVERVIEW OF FINAL STATUS SURVEY DESIGN

2.1 Introduction

It is recognized that demonstrating that residual concentrations of radioactivity at a site are at very low levels in the presence of background may be a complex task involving sophisticated sampling, measurement, and statistical analysis techniques. The difficulty of the task can vary substantially depending on a number of factors, including the radionuclides in question, the background level for those and other radionuclides at the site, and the temporal and spatial variations in background at or near the site. Sufficient radiological data must be collected to characterize both the residual radioactivity at the site and the background radioactivity levels in the vicinity of the site. The number of measurements required to accomplish this task will be determined on a site-specific basis and will depend upon the nature of the facility, its size, the selection of the statistical tests used, and certain statistical parameter values that influence how compliance with radiological criteria is determined.

2.2 Final Status Survey Design

Decommissioning is defined in 10 CFR 20.1003 as removing a facility or site safely from service, and reducing residual radioactivity to a level that permits (1) release of the property for unrestricted use and termination of the license; or (2) release of the property under restricted conditions and termination of the license. A *survey unit* is a geographical area of specified size and shape for which a separate decision will be made whether or not that area meets the release criteria. This decision is made following a *final status survey* of the survey unit. Thus, a survey unit is an area for which a final status survey is designed, conducted, and results in a release decision. The objective of this report is the design of efficient final status surveys. These surveys should obtain all of the data required for making the decision, but avoid the collection and analysis of superfluous samples.

Usually there are two conditions that would lead to the determination that a particular survey unit requires further remediation before unrestricted release:

(1) If the average level of residual radioactivity within the survey unit exceeds the regulatory limit, or

(2) If there are small areas within the survey unit with elevated residual radioactivity that exceed the regulatory limit.

Sampling at discrete points within the survey unit is a simple method for determining if the first of these conditions exists. The term *sampling* is used here in its statistical sense, namely obtaining data from a subset of a population. Sampling in this sense would include both direct *in situ* measurements and the collection of physical samples for laboratory analysis.

On the other hand, sampling at discrete points within a survey unit is not a very efficient method of determining if the second condition exists. Scanning is a much better method for detecting isolated areas with elevated activity. However, scanning is generally not as sensitive as sampling.

A major component of the survey designs discussed in this report is the efficient use of sampling at distinct locations combined with scanning to accurately determine the final status of a survey unit. The statistical procedures described in this report are used to establish the number of samples taken at distinct locations needed to determine if the mean concentration in the survey unit exceeds the regulatory limit, with a specified degree of precision. Thus, these statistical procedures are as important in the planning and design of the final status survey as they are in the analysis and interpretation of the resulting data.

2.2.1 Release Criteria

In the past, release criteria have often been expressed as activity concentration limits. The criteria for license termination given in 10 CFR 20.1402 and 10 CFR 20.1403 are expressed in terms of total effective dose equivalent (TEDE). This cannot be measured directly. *Exposure pathway modeling* is used to calculate the estimated volume or surface area concentration of specific radionuclides that could result in a TEDE equal to the release criterion. This concentration is termed the *derived concentration guideline level (DCGL)*. The units for the DCGL are the same as the units for measurements performed to demonstrate compliance (e.g., Bq/kg, Bq/m^2, etc.). This allows direct comparisons between the survey results and the DCGL.

A complete discussion of DCGLs is beyond the scope of this report. There is, however, one aspect of exposure pathway modeling that bears directly on survey unit design. That is the dependence of the TEDE on the assumed area of contamination used in the exposure pathway model.

The two conditions of Section 2.2 that may cause a survey unit to fail the TEDE release criterion may have very different corresponding DCGLs because of the different size of the areas of residual radioactivity. Consequently, this report considers two distinct DCGLs:

(1)   The $DCGL_W$ is derived assuming that residual radioactivity is uniformly distributed over a wide area, i.e. the entire survey unit. This can often be the default DCGL provided by an exposure pathway model.

(2)   The $DCGL_{EMC}$ is derived assuming that residual radioactivity is concentrated in a much smaller area, i.e., in only a small percentage of the entire survey unit.

The $DCGL_{EMC}$ can never be less than the $DCGL_W$, but it may be significantly greater. The ratio of the $DCGL_{EMC}$ to the $DCGL_W$ defines a radionuclide specific *area factor*, F_A, such that the $DCGL_{EMC} = (F_A) (DCGL_W)$, when the residual radioactivity is confined to an area of size A.

Detailed procedures for developing these area factors are beyond the scope of this report. However, in the simplest case, an area factor can be determined from the ratio of the result obtained from an exposure pathway model using the entire survey unit area to the result obtained assuming the residual radioactivity is confined to a smaller area. The value of the $DCGL_{EMC}$ that is calculated for survey planning purposes is based on an area, A, determined by the spacing between adjacent sampling locations.

2.2.2 Data Interpretation

The use of the two DCGLs discussed above differs when interpreting the results of the final status survey data. The $DCGL_w$ is used to form a statistical hypothesis concerning the level of residual radioactivity that may be uniformly distributed across the survey unit. A nonparametric test is applied to the sampling data taken at distinct locations in the survey unit to determine whether this level meets the release criterion.

The $DCGL_{EMC}$, however, is used to trigger further investigation of a portion of the survey unit. Any measurement from the survey unit is considered elevated if it exceeds the $DCGL_{EMC}$. This is the *elevated measurement comparison*. The existence of an elevated measurement in a survey unit indicates the possibility of an area of residual radioactivity that may cause the dose criteria to be exceeded. The elevated measurement alone does not indicate that the survey unit fails to meet the release criterion, only that it is a possibility that must be investigated further. The $DCGL_{EMC}$ is based on the area factor used for the survey design. The area factor used in the survey design is based on the area bounded by adjacent sampling points. The actual area of elevated activity could be smaller. Thus, the area factor based on the actual area of contamination may be larger. Further investigation will usually be necessary to determine the actual extent and concentration level of a specific elevated area.

2.2.3 Survey Unit Classification

To maximize the efficiency of the final status surveys, it is clear that the greatest effort should be expended on the areas that have the highest potential for contamination. Final status survey designs depend fundamentally on the *classification* of survey units according to contamination potential. The survey unit classification determines the final status survey design and the procedures used to develop the design.

Areas that have no potential for residual contamination are classified as *non-impacted areas*. These areas have no radiological impact from site operations and are typically identified early in decommissioning. Areas with some potential for residual contamination are classified as *impacted areas*.

Impacted areas are further divided into one of three classifications:

(1)   *Class 1 Areas*: Areas containing locations where, prior to remediation, the concentrations of residual radioactivity may have exceeded the $DCGL_w$.

(2)   *Class 2 Areas*: Areas containing no locations where, prior to remediation, the concentrations of residual radioactivity may have exceeded the $DCGL_w$.

(3)   *Class 3 Areas*: Areas with a low probability of containing any locations with residual radioactivity.

Class 1 areas have the greatest potential for contamination and therefore receive the highest degree of survey effort for the final status survey. Non-impacted areas do not receive any level of survey coverage because they have no potential for residual contamination. Impacted areas for which there is insufficient information to justify a lower classification should be classified as Class 1.

Examples of Class 1 areas include: (1) site areas previously subjected to remedial actions, (2) locations where leaks or spills are known to have occurred, (3) former burial or disposal sites, (4) waste storage sites, and (5) areas with contaminants in discrete solid pieces of material and high specific activity.

Remediated areas are identified as Class 1 areas because the remediation process often results in less than 100% removal of the contamination. The contamination that remains on the site after remediation is often associated with relatively small areas with elevated levels of residual radioactivity. This results in a non-uniform distribution of the radionuclide and a Class 1 classification. If an area is expected to have levels of residual radioactivity below the $DCGL_w$ and was remediated for purposes of ALARA, the remediated area might be classified as Class 2 for the final status survey.

Examples of areas that might be classified as Class 2 for the final status survey include: (1) locations where radioactive materials were present in an unsealed form, (2) potentially contaminated transport routes, (3) areas downwind from stack release points, (4) upper walls and ceilings of buildings or rooms subjected to airborne radioactivity, (5) areas handling low concentrations of radioactive materials, and (6) areas on the perimeter of former contamination control areas.

To justify changing the classification from Class 1 to Class 2, there should be measurement data that provides a high degree of confidence that no individual measurement would exceed the $DCGL_w$. Other justifications for reclassifying an area as Class 2 may be appropriate, based on site-specific considerations.

Examples of areas that might be classified as Class 3 include buffer zones around Class 1 or Class 2 areas, and areas with very low potential for residual contamination but insufficient information to justify a non-impacted classification.

The number of distinct sampling locations needed to determine if a uniform level of residual radioactivity within a survey unit exists does not depend on the survey unit size. However, the sampling density within a survey unit should reflect the potential for small elevated areas of residual radioactivity. Thus, the appropriate size for survey units formed within each of the three area classifications differs. Survey units with a higher potential for residual radioactivity should be smaller. Suggested maximum areas for survey units are:

Class 1 Structures .... 100 m² floor area
Class 1 Land areas .... 2,000 m²
Class 2 Structures .... 100 to 1,000 m²
Class 2 Land areas .... 2,000 to 10,000 m²
Class 3 Structures .... no limit
Class 3 Land areas .... no limit

The area of the survey unit should also be consistent with that assumed in the exposure pathway model used to calculate the $DCGL_W$. Survey units with structure surface areas less than 10 m² or land areas less than 100 m² may have unnecessarily high sampling densities, and should be avoided.

2.2.4 Final Status Survey Classification

Class 1 areas have the highest potential for containing small areas of elevated activity exceeding the release criterion. Consequently, both the number of sampling locations and the extent of scanning effort is the greatest. The final status survey is driven by the effort to provide reasonable assurance that if any areas with concentrations in excess of the $DCGL_{EMC}$ exist that then these areas will be found. Sampling is done on a systematic grid. The distance between sampling locations is made small enough that any elevated area that might be missed by sampling would be found by scanning. Scanning is performed over 100% of the survey unit. The minimum detectable concentration (MDC) of the scanning method must be lower than the $DCGL_{EMC}$.

Class 2 areas may contain residual radioactivity, but the potential for elevated areas is very small. Sampling is done on a systematic grid. The distance between samples is limited by limiting the maximum size of the survey unit. Scanning is performed systematically over the survey unit. Since Class 2 is an intermediate classification, scanning coverage may range from as little as 10% to nearly 100% of the survey unit, depending on whether the potential for an elevated area is nearer that for a Class 1 area or for a Class 3 area.

Class 3 areas should contain little, if any, residual radioactivity. There should be virtually no potential for elevated areas. Sampling is random across the survey unit, and the sample density can be very low. Scanning is limited to those parts of the survey unit where it is deemed prudent, based on the judgment of an experienced professional.

Table 2.1 summarizes the differences in the final status survey design for each of the three survey unit classifications.

Table 2.1 Final Status Survey Design Classification

Class	Sampling	Scanning
1	Systematic	100% Coverage
2	Systematic	10 – 100%
3	Random	Judgmental

2.2.5 Background

The release criteria in 10 CFR Part 20.1402 and 1403 specify a dose limit (TEDE) due to residual radioactivity that is *distinguishable from background radiation.* According to 10 CFR 20.1003, *background radiation* means radiation from cosmic sources, naturally occurring radioactive material, including radon (except as a decay product of source or special nuclear material), and global fallout as it exists in the environment from the testing of nuclear explosive devices or from nuclear accidents like Chernobyl which contribute to background radiation and are not under the control of the licensee. Background radiation does not include radiation from source, byproduct, or special nuclear materials regulated by the Commission. The term *distinguishable from background* means that the detectable concentration of a radionuclide is statistically different from the background concentration of that radionuclide in the vicinity of the site, or, in the case of structures, in similar materials using adequate measurement technology, survey and statistical techniques.

For the purposes of survey design, the method of accounting for background radiation will depend not only on the radionuclides involved, but also on the type of measurements made. For radionuclide specific measurements of radionuclides that do not appear in natural background, it is clear that no adjustments for background are needed. In some cases, a sample-specific background adjustment may be possible. For example, residual ^{238}U activity may be distinguishable from natural ^{238}U by the amount of ^{226}Ra present in a sample. In other cases, it will not be possible to make such a distinction. In particular, such a distinction will not be possible, even if the radionuclide does not appear in background, when gross activity or exposure rate measurements are used.

For the elevated measurement comparison of individual sampling results, an adjustment for background will not ordinarily be necessary, since the DCGL$_{EMC}$ is a multiple of the DCGL$_w$. For statistical testing of the results against the release criterion, however, one approach is used when the measurements represent net residual radioactivity, but a different approach is necessary when the measurements represent total radioactivity including background.

When a specific background can be established for individual samples, the results of the survey unit measurements can be compared directly to the DCGL, since each is a measurement of the residual radioactivity alone. Because only one set of measurements is involved in this comparison, the statistical test is called a *one-sample test.*

When a specific background cannot be established for individual samples, the survey unit measurements cannot be directly compared to the DCGL, since each is a measurement of the total of any residual radioactivity plus the survey unit background. In this case, the measurements in a survey unit must be compared to similar measurements in local *reference areas* that have been matched to the survey unit in terms of geological, chemical, and biological attributes, but which have not been affected by site operations. The *distribution* of the measurements in a survey unit is compared to the *distribution* of background measurements in a reference areas. Because two sets of measurements are used in making this comparison, the statistical test is called a *two-sample test.*

2.2.6 Data Variability

The ease or difficulty with which compliance may be demonstrated depends primarily on the size of the $DCGL_W$ relative to the amount of variability in the measurement data. This is commonly known as the signal-to-noise ratio. As this ratio becomes smaller, more measurement data will be needed to determine compliance with the release criterion, i.e. to extract the signal from the noise.

The variability in the measurement data is a combination of the precision of the measurement process, and the real spatial variability of the quantity being measured in the survey unit. Variability can be reduced by using more precise measurement methods, but the spatial variability remains. The mechanism by which spatial variability can be reduced is by choosing survey units that are as homogeneous as possible with respect to the expected level of residual radioactivity. This means that survey units should generally be formed from areas with similar construction, use, contamination potential, and remediation history.

If the measurement data include a background contribution, the spatial variability of background adds to the overall measurement variability. Thus, the survey units where such measurements will be used should be as homogeneous as possible with respect to expected natural background as well. Further information on natural background and its variability can be found in NUREG-1501 (August 1994).

An additional source of variability is introduced when survey unit measurement data including background are compared to measurements from a reference area. Any systematic difference in background level between the survey unit and the reference area will be indistinguishable from a difference in residual radioactivity in the two areas. This situation is not unique to decommissioning or the methodology of this report. It is always true when a background adjustment must be made using data from a location other than the sampling location, e.g. using control dosimeters at remote locations to account for background in monitoring dosimeters.

2.2.7 Reference Areas

A *reference area* (or background area) is a geographical area from which representative samples of background will be selected for comparison with samples collected in specific survey units at the remediated site. The reference area should have similar physical, chemical, radiological, and biological characteristics to the site area being remediated, but should not have been contaminated by site activities. The reference area is where background is measured and defined for the purpose of decommissioning. To minimize systematic biases in the comparison, the same sampling procedure, measurement techniques, and type of instrumentation should be used at both the survey unit and the reference area. The distribution of background measurements in the reference area should be the same as that which would be expected in the survey unit if that survey unit had never been contaminated. It may be necessary to select more than one reference area for a specific site, if the site includes so much physical, chemical, radiological, or biological variability that it cannot be represented by a single reference background area.

2.2.8 Radionuclide-Specific Measurements

As indicated in Section 2.2.5 and 2.2.6, if radionuclide-specific survey methods are used, and if the radionuclide of interest does not appear in background, reference area measurements are not needed, and one-sample statistical tests are used. If other survey methods are used, such as gross activity or exposure rate measurements, then the individual contributions due to background and any residual radioactivity will not be separately identifiable, suitable reference area measurements will be needed, and two-sample statistical tests are used.

Even if the radionuclide of interest does appear in background, the variability of radionuclide specific measurements will generally be smaller than those of gross activity measurements in the same area. Depending on the level of residual activity that it is necessary to detect, many more measurements may be required if gross activity or exposure rate measurements are used than if radionuclide-specific measurements are made. At very low levels, it may be difficult or impossible to distinguish the residual radioactivity contribution unless radionuclide-specific methods are used. However, it may be economical in some circumstances to perform a larger number of simpler, less expensive measurements. One of the primary advantages of the Data Quality Objectives process, is that alternative measurement strategies can be compared at the planning stage. Exploring the statistical design of the final status survey in advance, the most efficient method for the problem can be chosen.

2.3 Statistical Concepts

This section introduces some of the statistical concepts and terminology used in hypothesis testing. A use of statistical hypothesis testing that is familiar in the radiation protection measurements field is the calculation of lower limits of detection. The methodology of this report can be viewed as an application of these same concepts to a survey unit rather than to a laboratory measurement. This analogy is pursued further in Section 2.6.

2.3.1 Null and Alternative Hypotheses

The decisions necessary to determine compliance with the criteria for license termination are formulated into precise statistical statements called hypotheses. The truth of these hypotheses can be tested with the survey unit data. The state that is presumed to exist in reality is expressed as the null hypothesis (denoted by H_0). For a given null hypothesis, there is a specified alternative hypothesis (denoted as H_a), which is an expression of what is believed to be the state of reality if the null hypothesis is not true.

For the purposes of this report, the important decision is whether or not a site meets the applicable license termination and release criteria. This decision will be supported by the individual decisions on whether each survey unit meets the applicable release criteria. In this report, two different scenarios, designated Scenario A and Scenario B, are considered.

In Scenario A, the null hypothesis is:
H_0: The survey unit does not meet the release criterion
   versus the alternative
H_a: The survey unit meets the release criterion.

In Scenario B, the null hypothesis is:

H_0: The survey unit meets the release criterion.

versus the alternative

H_a: The survey unit does not meet the release criterion.

As indicated in Section 2.2.1, the release criterion is specified in terms of a dose, which is converted via pathway modeling to a residual radioactivity concentration limit, the $DCGL_W$. If the concentration of residual radioactivity that is distinguishable from background in the survey unit exceeds the $DCGL_W$, the survey unit does not meet the release criterion.

When choosing the scenario to use, it is important to note that the null hypothesis cannot be proved, i.e. accepted as true. The null hypothesis is either rejected or not rejected. The data are either consistent with the null hypothesis, or they are not. It is stated this way because there are two circumstances leading to the decision not to reject the null hypothesis:

(1) the null hypothesis is true.

(2) the null hypothesis is false, but the data did not provide enough evidence to show it.

The burden of proof is on the alternative. Thus, in Scenario A, the survey unit will not be released until proven clean. In Scenario B, the survey unit will be released unless it is shown to be contaminated above background. Rejecting the null hypothesis has different implications for survey unit release in the two scenarios. For this reason, a survey unit will be said to *pass* the final status survey if it is concluded that it may be released. Otherwise it will be said to *fail*. In Scenario A, the emphasis is on the dose limit. In Scenario B, the emphasis is on indistinguishability from background. In Scenario A, the survey unit is assumed to fail unless the data show it may be released. In Scenario B, the survey unit is assumed to pass unless the data show that further remediation is necessary.

In Scenario A, the measured average concentration in the survey unit must be significantly less than the $DCGL_W$ in order to pass. In Scenario B, the measured average concentration in the survey unit must be significantly greater than background in order to fail. In Scenario A, increasing the number of measurements in a survey unit increases the probability that an adequately remediated survey unit will pass. In Scenario B, increasing the number of measurements in a survey unit increases the probability that an inadequately remediated survey unit will fail.

Which scenario should be used? Because of insufficient evidence, the null hypothesis may not be rejected even when it is false. Thus, the null hypothesis should be the one that is the easiest to live with even if it is false. The alternative should be the hypothesis that carries the severest consequences if it falsely chosen. To make the proper choice of scenario, the possible types of decision errors and the probability of making them should be examined. This is the subject of the next section.

In most cases, when the $DCGL_W$ is fairly large compared to the measurement variability, Scenario A should be chosen. This is because even contamination below the $DCGL_W$ should be measurable. Requiring additional remediation when it is not strictly necessary may still have some benefit in the form of reduced radiation exposure. Releasing a survey unit that really should be remediated further is a less tolerable mistake. It is anticipated that Scenario A will be simpler to implement for most licensees.

When the $DCGL_W$ is small compared to measurement and/or background variability, Scenario B should be chosen. This is because contamination below the $DCGL_W$ will be difficult to measure. Requiring additional remediation when it is not necessary, may essentially require remediation of background. This is an impossible task. Releasing a survey unit that has residual radioactivity within the range of background variations is a less severe consequence in this case. It is fairly straightforward to specify what is meant for a survey unit to meet the release criterion, but a survey unit may be distinguishable from background either because it is uniformly contaminated or because it contains spotty areas of residual radioactivity. For this reason, the data analysis for Scenario B involves two statistical tests performed in tandem.

In this report, the two scenarios are developed in parallel. Within the limits imposed by the magnitude of the data variability relative to the $DCGL_W$, essentially the same information about the survey unit should be obtained, and the same conclusion regarding compliance should be reached using either scenario. The difference is in the emphasis.

2.3.2 Decision Errors

Errors can be made when making site remediation decisions. The use of statistical methods allows for controlling the probability of making decision errors. When designing a statistical test, acceptable error rates for incorrectly determining that a site meets or does not meet the applicable decommissioning criteria must be specified. In determining these error rates, consideration should be given to the number of sample data points that are necessary to achieve them. Lower error rates require more measurements, but result in statistical tests of greater power and higher levels of confidence in the decisions. In setting error rates, it is important to balance the *consequences* of making a decision error against the *cost* of achieving greater certainty.

There are two types of decision errors that can be made when performing the statistical tests described in this report. The first type of decision error, called a Type I error, occurs when the null hypothesis is rejected when it is actually true. A Type I error is sometimes called a "false positive." The probability of a Type I error is usually denoted by α. The Type I error rate is often referred to as the significance level or size of the test.

The second type of decision error, called a Type II error, occurs when the null hypothesis is not rejected when it is actually false. A Type II error is sometimes called a "false negative." The probability of a Type II error is usually denoted by β. The *power* of a statistical test is defined as the probability of rejecting the null hypotheses when it is false. It is numerically equal to $1-\beta$, where β is the Type II error rate.

The setting of acceptable error rates is a crucial step in the planning process. Specific considerations for establishing these error rates are discussed in Chapter 3. Table 2.2 summarizes the types of decision errors that can be made for the specific hypotheses of Scenario A and Scenario B.

Table 2.2 Summary of Types of Decision Errors

Scenario A	True Condition of Survey Unit	
Decision Based on Survey	**Does Not Meet Release Criterion**	**Meets Release Criterion**
Does Not Meet Release Criterion	*Survey unit fails* Correct Decision (Probability = $1-\alpha$)	*Survey unit fails* Type II Error (Probability = β)
Meets Release Criterion	*Survey unit passes* Type I Error (Probability = α)	*Survey unit passes* Correct Decision (Power = $1-\beta$)

Scenario B	True Condition of Survey Unit	
Decision Based on Survey	**Meets Release Criterion**	**Does Not Meet Release Criterion**
Meets Release Criterion	*Survey unit passes* Correct Decision (Probability = $1-\alpha$)	*Survey unit passes* Type II Error (Probability = β)
Does Not Meet Release Criterion	*Survey unit fails* Type I Error (Probability = α)	*Survey unit fails* Correct Decision (Power = $1-\beta$)

2.4 Hypothesis Testing Example

The following example illustrates the use of the concepts discussed above as currently used in the determination of detection limits for radioactivity measurements. The analogy is most direct for Scenario B.[1] The calculation of detection limits, which is generally familiar to radiation protection professionals, also involves hypothesis testing (HPSR/ EPA 520/1-80-012, 1980; NUREG/CR-4007, 1984; Currie, 1968). In this situation, there is a measurement error, often taken to be the Poisson counting error, σ, equal to the square root of the number of counts. There is a background counting rate, and any additional radioactivity in a sample must be distinguishable above that. Generally it is assumed that the number of counts is sufficiently large so that a normal approximation to the Poisson distribution of counts is appropriate.

---

[1] For Scenario A, the analogy would have to be restructured for the problem of deciding whether a given sample, assumed to contain added radioactivity exceeding L_D actually contained a smaller amount. In essence, the null and alternative hypotheses would be reversed.

2.4.1 Detection Limits

For the calculation of detection limits, the hypotheses are:

Null Hypothesis:
H_0: The sample contains no radioactivity above background.
 and
Alternative Hypothesis:
H_a: The sample contains added radioactivity.

The count obtained from the sample measurement is the test statistic, and it has a different probability distribution under the null and alternative hypothesis (see Figure 2.1). If a sample that contains no radioactivity above background is declared to contain radioactivity above background, a Type I error is made. Conversely, if a sample that contains radioactivity above background is declared to contain no radioactivity above background, a Type II error is made.

The Type I error rate, α, depends on the variability of background, i.e., it is controlled by requiring that the net counts exceed a certain multiple of the measurement standard deviation. Under the null hypothesis, namely when there is no radioactivity above background, the net counts have mean $B - B = 0$.

The standard deviation of the net count is

$$\sigma_{B-B} = \sqrt{B + B} = \sqrt{\sigma^2 + \sigma^2} = \sqrt{2}\sigma \qquad (2\text{-}1)$$

where B is the background count, and $\sigma = \sqrt{B}$ is its standard deviation, since for a Poisson distribution the standard deviation is the square root of the mean. Unless the mean number of counts is very low, a normal distribution with the same mean and standard deviation can be used to approximate the Poisson distribution of the background counts. This determines the critical level, L_C. If a net count above the critical detection level is obtained, the null hypothesis is rejected. That is, the decision is made that the sample being measured contains radioactivity above background.

$$L_C = z_{1-\alpha}\, \sigma_{B-B} = z_{1-\alpha}\, \sqrt{2}\sigma \qquad (2\text{-}2)$$

$z_{1-\alpha}$ is the $1-\alpha$ percentile of a standard normal distribution, e.g. if $\alpha = 0.05$, then $1 - \alpha = 0.95$ and $z_{1-\alpha} = 1.645$. Note that the distribution of background counts (lefthand curve in Figure 2.1) is used for this calculation.

The Type II error rate, β, depends on the variability of the added radioactivity and is controlled by requiring that the net counts exceed a certain number of standard deviations above the critical level.

$$L_D = L_C + z_{1-\beta}\, \sigma_{L_D} = z_{1-\alpha}\, \sqrt{2}\sigma + z_{1-\beta}\, \sigma_{L_D} = z_{1-\alpha}\, \sqrt{2}\sigma + z_{1-\beta} \sqrt{L_D + 2\sigma^2} \qquad (2\text{-}3)$$

since

$$\sigma_{L_D} = \sqrt{(L_D+B)+B} = \sqrt{L_D+2\sigma^2}$$

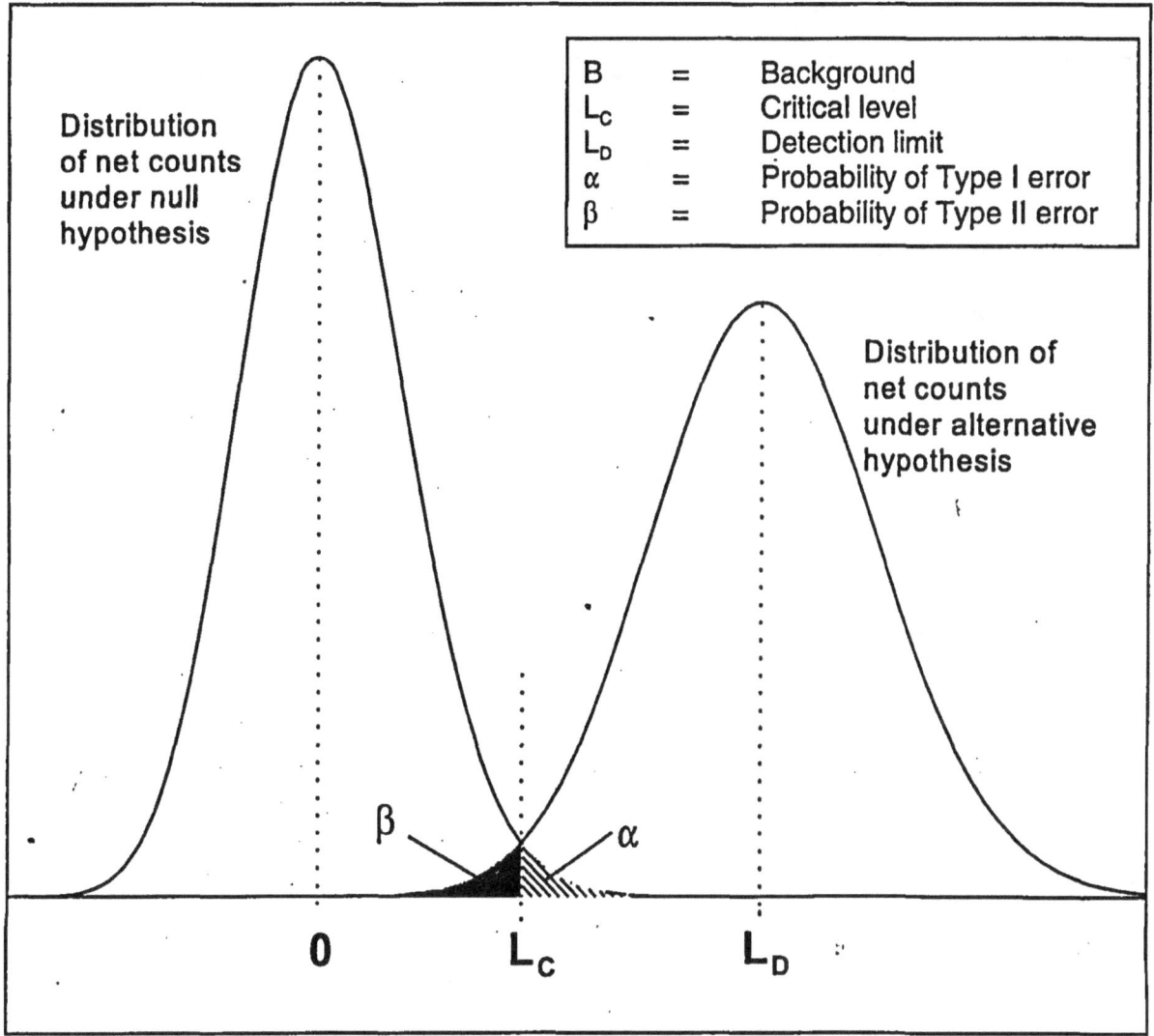

B	=	Background
L_C	=	Critical level
L_D	=	Detection limit
α	=	Probability of Type I error
β	=	Probability of Type II error

Figure 2.1 Type I and Type II Errors in the Determination of a Detection Limit

The distribution of counts under the alternative hypothesis (right hand curve in Figure 2.1) is used to derive Equation 2-3. If the probability of a Type II error is set the same as the probability of a Type I error, then $z_{1-\alpha} = z_{1-\beta} = k$. Solving Equation 2-3 for L_D, the count detection limit is found to be

$$L_D = k^2 + 2k\sqrt{2}\,\sigma = k^2 + 2\,L_C \qquad (2\text{-}4)$$

The power, $1-\beta$, is the probability that the measurement will indicate the presence of additional radioactivity in the sample, when the sample actually contains additional activity in the amount necessary to produce an average of L_D counts above background during the measurement.

NUREG-1505

2.4.2 Final Status Surveys

The statistical procedures described in this report for final status surveys have many similarities to the detection limit calculation. Corresponding to Figure 2.1, the relationship between the hypothesis, decision error rates and measurement distributions in Scenario A and Scenario B are shown in Figures 2.2 and 2.3, respectively.

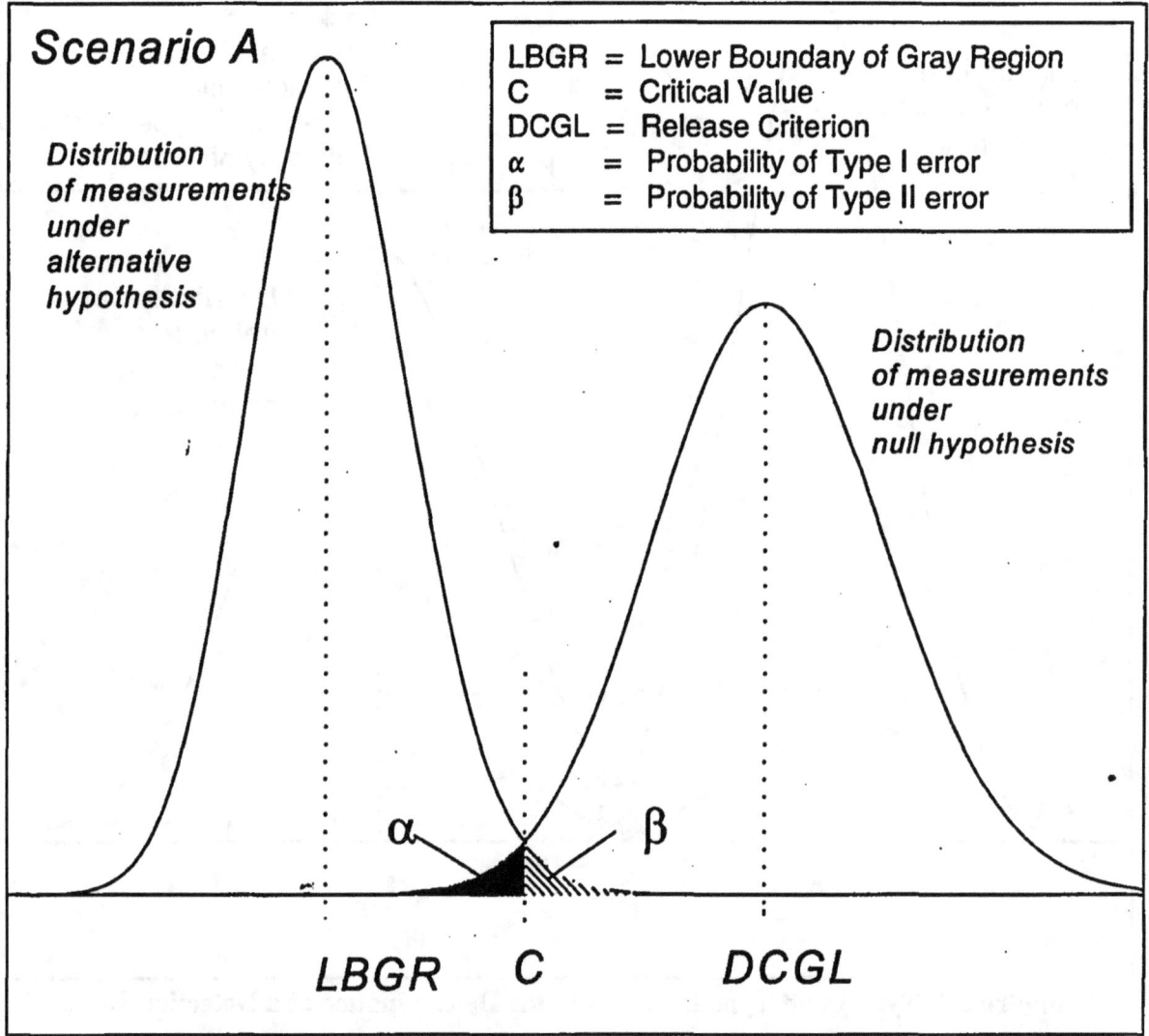

Figure 2.2 Type I and Type II Errors for Scenario A

Some other points of similarity are:

(1) The null hypothesis is:
    H_0: The sample contains no radioactivity above background.
      becomes either
    H_0: The survey unit does not meet the release criterion (Scenario A).
      or
    H_0: The survey unit meets the release criterion (Scenario B).

(2) The alternative hypothesis is:

H$_a$: The sample contains added radioactivity above the detection limit.
becomes either

H$_a$: The survey unit meets the release criterion (Scenario A).
or

H$_a$: The survey unit does not meet the release criterion (Scenario B).

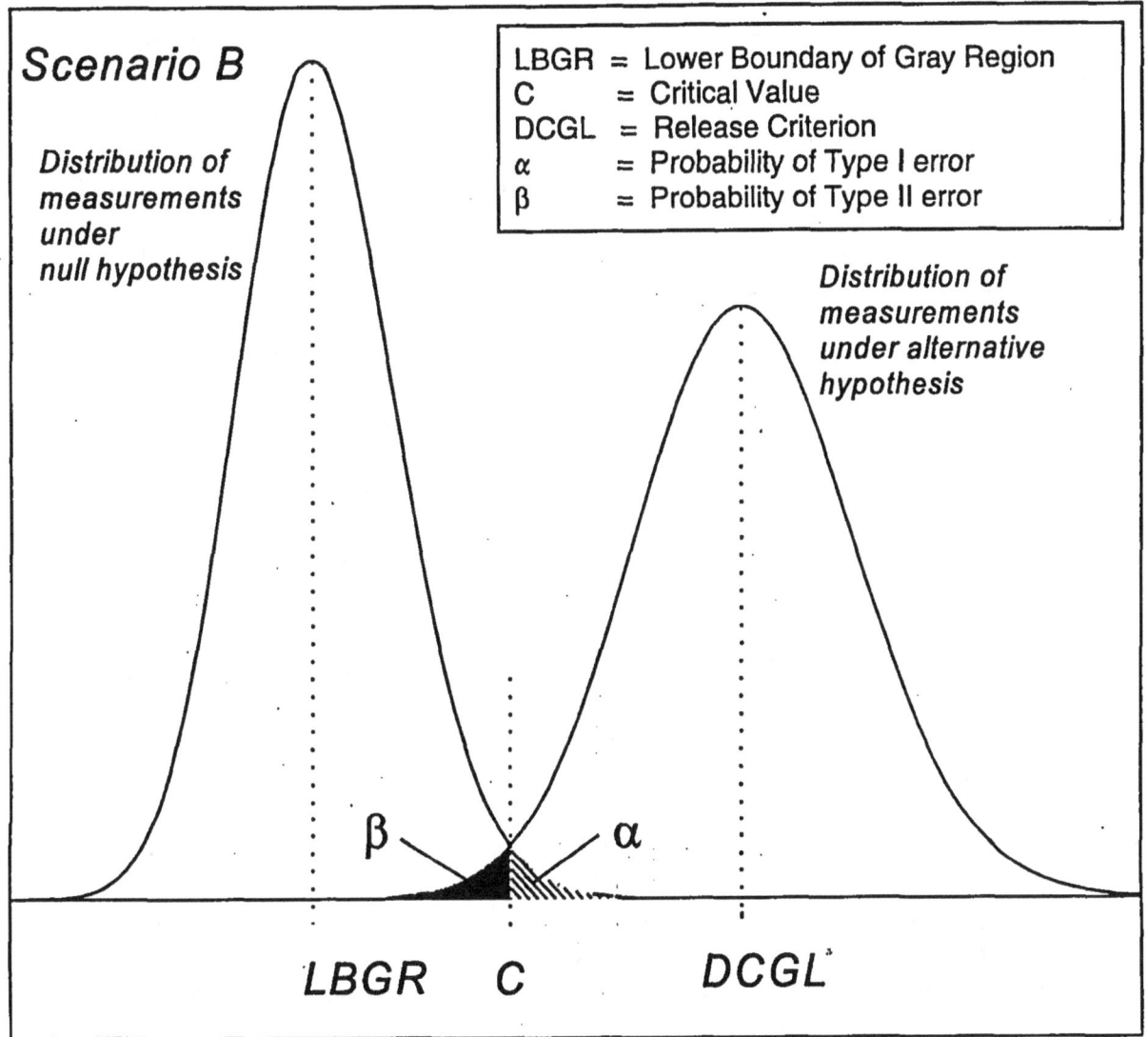

Figure 2.3 Type I and Type II Errors for Scenario B

(3) The Type I error rate is computed using the distribution of counts assuming the null hypothesis is true. Similarly, the Type I error rates for the tests described in this report will be calculated using the distribution of the measurements under the null hypothesis.

(4) The Type II error rate is computed using the distribution of counts assuming the alternative hypothesis is true. Similarly, the Type II error rates for the tests described in this report will

be calculated using the distribution of the measurements under the alternative hypothesis. This also gives the power of the tests.

(5) The variability of the count obtained from the sample, σ, plays a crucial role in determining the value of the detection limit. Similarly, the variability of the radioactivity measurements in the reference areas and survey units plays a crucial role in how well the tests described in this report will perform.

(6) Corresponding to the detection limit , L_D, is a critical level of counts, L_C. Any sample producing more than the critical level of counts is assumed to contain additional radioactivity. Thus, the decision whether or not to reject the null hypothesis is based on comparing the counts actually obtained from the sample to the critical detection level. Similarly, the decision whether or not to reject the null hypothesis for a survey unit is based on the critical level of a test statistic which is computed from the measurement data. Note that while L_C and L_D are expressed in counts, there is a corresponding concentration level in the sample being measured that will, on average, give rise to that number of counts.

(7) The critical level of counts, L_C is calculated so that the decision to reject the null hypothesis is made with probability α when the true concentration in the sample being measured is zero. The critical value of the final status survey test statistic is calculated so that the decision to reject the null hypothesis in Scenario B is made with probability α when the true concentration is equal to a certain value called the LBGR (Lower Boundary of the Gray Region). The LBGR is a concentration value between zero and the $DCGL_W$ at which probability of the survey unit incorrectly failing the final status survey is specified. The LBGR is discussed further in Section 3.7.

(8) The detection limit can usually be made lower by counting for a longer time, thereby reducing the relative measurement error, at additional cost. Similarly, the ability of the tests described in this report to distinguish smaller amounts of residual radioactivity from background more accurately can be improved by taking a greater number of samples, at additional cost.

(9) Usually, a detection limit is calculated given the Type I and Type II error rates and the background variability. However, if a certain detection limit is pre-specified instead, the procedure given above shows how to relate it to the Type I and Type II error rates, and the measurement variability. Similarly, the procedures of this report will show the interrelationship of the decommissioning criteria (dose above background), the Type I and Type II error rates, and the measurement variability.

2.4.3 The Effect of Measurement Variability on the Decisionmaking Process

Figure 2.4 further illustrates the affect of the measurement standard deviation on the decision process. Shown are three hypothetical measurement distributions, with true mean concentration , equal to zero, one, and three times the measurement standard deviation, σ. Assume for simplicity there is no background to subtract. Then the critical level, $L_C = z_{1-\alpha}\sigma = 1.645\sigma$ when $\alpha = 0.05$. Thus, there is a 5% chance of a positive result when the true concentration is actually zero. If the true concentration is 3σ, the probability of a positive result is very high since most of the distribution lies above L_C (91% using the normal distribution table with $z = 3 - 1.645 = 1.355$).

However, if the true concentration is 1σ, then there is less than a 50% chance of a positive result (26% using the normal distribution table with $z = 1-1.645 = -0.645$). If a true mean concentration of $C = 1\sigma$ must be measured, then the uncertainty must be reduced by taking more measurements. If nine measurements are averaged, then the standard deviation of the mean, σ^*, falls by a factor of three (one over the square root of the number of measurements). In the "new standard deviation units" $C = 1\sigma = 3\sigma^*$. Thus, a difference of 1σ can be distinguished with nine measurements as easily as a difference of 3σ can be distinguished with one measurement.

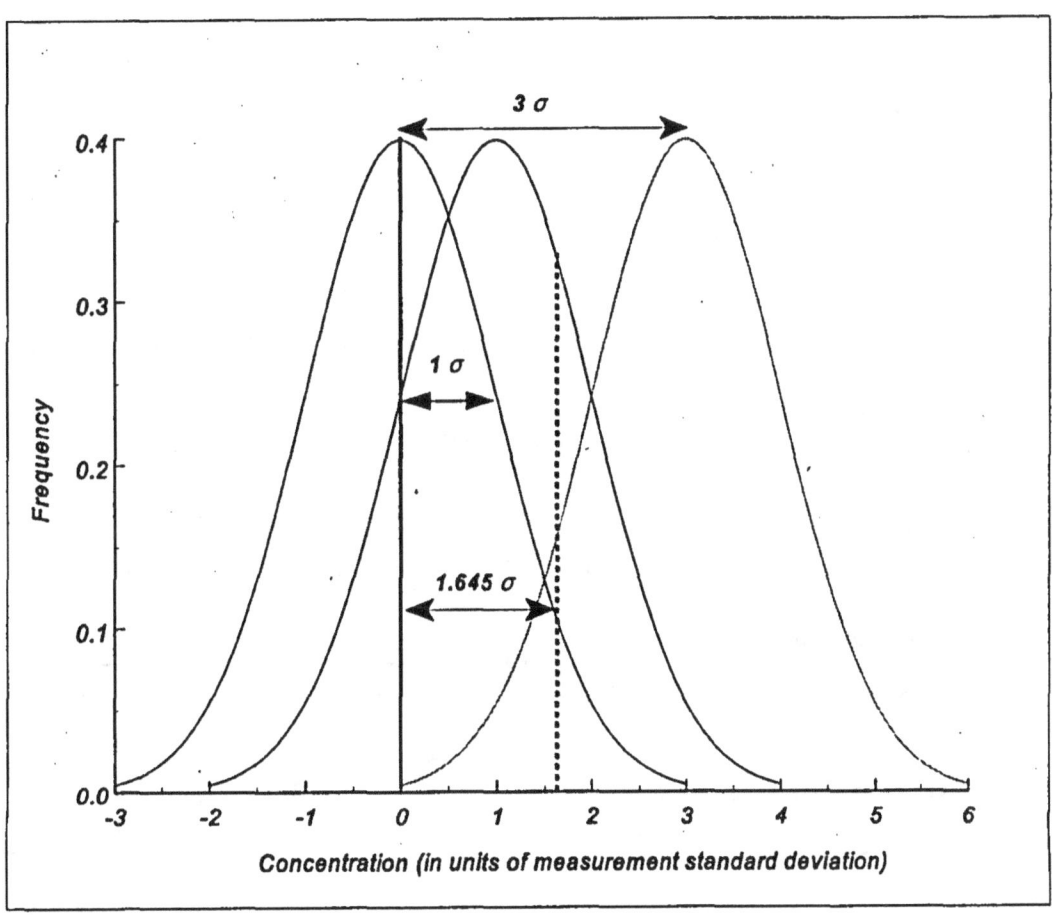

Figure 2.4 Differences in Concentration Compared to Measurement Variability

2.5 Statistical Tests

There are two important uses of the statistical tests described in this report. The first is in the analysis of the final status survey data to demonstrate compliance with the release criterion. However, the second, and perhaps more important use, is in the design of the final status survey. In some cases it may be clear from the data, without any formal analysis, whether or not a survey unit meets the decommissioning criteria. Provided that an adequate number of measurements are made (either *in situ* or from samples), Table 2.3 can be used to determine whether or not a formal statistical test is necessary.

NUREG-1505

It is usually not obvious, *a priori,* what number of samples is necessary in order to show whether or not a survey unit meets the decommissioning criteria with acceptable rates of error. The DQO process described in Chapter 3 provides a general method for designing surveys so that accurate remediation decisions can be made cost effectively. Simple estimates of the number of samples required for the statistical tests may be made using the mathematical relationships between the error rates, residual radioactivity levels, and measurement variability.

Table 2.3 Summary of Statistical Tests

Radionuclide not in background *and* radionuclide-specific measurements made	
Survey Result	**Conclusion**
All measurements below $DCGL_W$	Survey unit meets release criterion
Average above $DCGL_W$	Survey unit does not meet release criterion
Otherwise (some measurements above $DCGL_W$, but average below $DCGL_W$)	Conduct Sign test and elevated measurement comparison

Radionuclide in background *or* non-radionuclide-specific measurements made	
Survey Result	**Conclusion**
Difference between maximum survey unit measurement and minimum reference area measurements is below $DCGL_W$	Survey unit meets release criterion
Difference of survey unit average and reference area average is above $DCGL_W$	Survey unit does not meet release criterion
Otherwise (Maximum difference above $DCGL_W$, but average difference below $DCGL_W$)	Conduct WRS test and elevated measurement comparison

2.5.1 Nonparametric Statistical Tests

Many statistical tests can be used for determining whether or not a survey unit meets the release criteria. Any one test may perform better or worse than others, depending on the hypotheses to be tested, i.e., the decision that is to be made and the alternative, and how well the assumptions of the test fit the situation.

The basic distinction between parametric and nonparametric statistical techniques is that parametric techniques use specific assumptions about the probability distributions of the measurement data. The most commonly made assumption is that the data fit a normal distribution. Such is the case when the Student's t statistic is used. Additional data and statistical tests would generally be necessary in order to show that the assumption of normality is justified (EPA QA/G-9), 1995.

Nonparametric techniques (sometimes referred to as distribution-free statistical methods) can be used without assuming a particular underlying distribution. Thus, nonparametric techniques are often more appropriate in situations when the probability distribution of the data is either unknown or is some continuous distribution other than the normal distribution. That a statistical approach is nonparametric or distribution free does not imply that it is free of any and all assumptions about the data distribution. Most nonparametric procedures require that measured values be independent and identically distributed. Some nonparametric procedures assume that the underlying probability distribution is symmetric. However, these requirements are usually less restrictive than the assumption that the data follow a particular symmetric distribution, such as the normal distribution.

Parametric methods rely on the assumptions about the data distribution to infer how large the difference between two measurements is expected to be. These methods are better only if the assumptions are true. Many nonparametric techniques are based on ranking the measurement data. The data are ordered from smallest to largest, and assigned numbers (ranks) 1, 2, 3,... accordingly. The analysis is then performed on the *ranks* rather than on the original measurement values. The advantage of this approach is that the probability that one measurement is larger (i.e. ranked higher) than another can be computed exactly by combinatorial (enumeration and counting) methods without reference to a specific probability distribution.

If the underlying distribution is known, a parametric test can make use of that additional information. If the underlying distribution is different than that assumed, however, the results can be unpredictable. The nonparametric methods described in this report have been found to perform nearly as well as the corresponding parametric tests, even when the conditions necessary for applying the parametric tests are fulfilled. There is often relatively little to be gained in efficiency from using a specific parametric procedure, but potentially much to be lost. Thus, it may be considered prudent to use nonparametric methods in all cases.

For survey measurements at or near background, there may be some measurement data which are at or below instrumental detection limits. These data are sometimes reported as "less than" or "non-detects". Such data are not easily treated using parametric methods. It is recommended that the actual numerical results of measurements always be reported, even if these are negative or below calculated detection limits. However, if it is necessary to analyze data which include non-numerical results, nonparametric procedures based on ranks can still be used in many cases. This is an additional advantage to the use of these methods.

2.5.2 Wilcoxon Rank Sum and Sign Tests

The Wilcoxon Rank Sum (WRS) test and Sign test are used in this report to detect a uniform shift in the mean of a distribution of measurements. Without assuming symmetry in the measurement distribution, these tests are technically for the median. However, computer simulations have shown that these tests generally produce the correct decisions more often when the assumption of symmetry is violated than the commonly used Student's t-test, which assumes normality in addition to symmetry. Nevertheless, extremes of asymmetry are guarded against by conducting the elevated measurement comparison in addition to the WRS and Sign tests. This issue is discussed further in the next section.

The WRS test is a two-sample test that compares the distribution of a set of measurements in a survey unit to that of a set of measurements in a reference area. The Sign test is a one-sample test that compares the distribution of a set of measurements in a survey unit to a fixed value, namely the derived concentration limit for a specific radionuclide.

The WRS test, also known as the Mann-Whitney test, is performed for Scenario A by first adding the value of the DCGL to each of the reference area measurements, and then listing the *combined* set of survey unit and adjusted reference area measurements in increasing numerical order from smallest to largest. The next step is to replace the measurement values by their ranks, i.e., their position number in the ordered list. Thus, the ranks are simply integer values from 1 through N, where N is the total number of combined measurements. The rank 1 is assigned to the smallest value, 2 to the second smallest observation, etc. Then, the sum of the ranks of the survey unit and the sum of the ranks of the adjusted reference area measurements is computed. Because the sum of the combined ranks is a fixed constant equal to $N(N+1)/2$, the sum of the adjusted reference area measurement ranks is equal to $N(N+1)/2$ minus the sum of the ranks of the survey unit measurements.

If the level of residual radioactivity in the survey unit is exactly at the DCGL, then any given rank is equally likely to belong to either an adjusted reference area measurement or a survey unit measurement. If the survey unit has residual radioactivity less than the DCGL, the survey unit site ranks will tend to be smaller than the adjusted reference area ranks. The larger the average of the ranks of the adjusted reference area measurements relative to the ranks of the survey unit measurements, the smaller the probability that it is by chance, and the greater the evidence that the residual radioactivity in the survey unit is actually below the DCGL. If the sum of the ranks of the adjusted reference area measurements exceeds a calculated critical value, the decision is made to reject the null hypothesis and to conclude that the survey unit actually meets the release criterion. In some cases, the result will be obvious without any computations. If, for example, all of the survey unit measurements are less than the smallest of the adjusted reference area measurements, then the sum of the ranks of the adjusted reference area measurements is at its maximum possible value, and the null hypothesis will always be rejected.

For Scenario B, the WRS is performed by first subtracting the value of the LBGR from each of the survey unit measurements, and then listing the combined set of adjusted survey unit and reference area measurements in increasing numerical order from smallest to largest and finding their ranks. As above, the sum of the ranks of the survey unit measurements and the sum of the ranks of the unadjusted reference area measurements is computed. Again, the sum of the reference area measurement ranks plus the sum of the ranks of the adjusted survey unit measurements is equal to $N(N+1)/2$. If the level of residual radioactivity in the survey unit is exactly at the LBGR, then any given rank is equally likely to belong to either a reference area measurement or a survey unit measurement. Thus, there is no reason to believe that the average of the survey unit ranks will differ greatly from the average of the reference area ranks. With residual radioactivity at the LBGR, the probability that the average of the survey unit ranks will be larger than the average of the reference area ranks is 50% by random chance. However, the larger the average of the survey unit ranks, the smaller the probability that it is by chance, and the greater the evidence that the survey unit is contaminated. If the sum of the survey unit ranks exceeds a calculated critical value, one can decide that the evidence shows that the residual radioactivity in the survey unit exceeds the LBGR.

The one-sample Sign test is performed for Scenario A by first subtracting each survey unit measurement from the derived concentration limit. Then the *number* of *positive* differences is counted. Large numbers of positive differences are evidence that the survey unit measurements do not exceed the derived concentration guideline.

The Sign test is performed for Scenario B by first subtracting the LBGR from each survey unit measurement. Then the *number* of *positive* differences is counted. Large numbers of positive differences are evidence that the survey unit measurements exceed the LBGR.

The Sign test uses no assumptions about the shape of the data distribution. An alternative test, the Wilcoxon Signed Rank (WSR) test, assumes that the measurement distribution is symmetric. When this assumption is valid, the WSR test generally has greater power than the Sign test. The WSR and other alternative statistical tests are discussed in Section 14.1.

2.5.3 Mean and Median

The WRS and Sign tests are designed to determine whether or not a degree of residual radioactivity remains uniformly throughout the survey unit. Since these methods are based on ranks, the results are generally expressed in terms of the median. When the underlying measurement distribution is symmetric, the mean is equal to the median. The assumption of symmetry is less restrictive than that of normality, since the normal distribution is itself symmetric. If, however, the measurement distribution is skewed to the right, the average will generally be greater than the median. In severe cases, it may happen that the average exceeds the $DCGL_W$ while the median does not. This is why the average is used to screen the data set before any statistical test is performed (see Table 2.3).

Figure 2.5 illustrates the potential differences between the median and the mean. The normal and lognormal distributions are commonly used examples of symmetric and skewed distributions, respectively. In this figure, the probability densities all have arithmetic mean equal to one. The coefficient of variation (arithmetic standard deviation divided by the mean) varies from 0.1 to 1.0. For values of the coefficient of variation larger than about 0.3, the lognormal begins to diverge significantly from the normal. When the coefficient of variation is 1.0, the difference between the median and the mean is large.

When the underlying data distribution is highly skewed, it is often because there are a few high measurements. Since the elevated measurement comparison is used to detect such measurements, the difference between using the median and the mean as a measure for the degree to which uniform residual radioactivity remains in a survey unit tends to diminish in importance. This is especially true in Scenario A, where the null hypothesis is that the survey unit does not meet the release criterion.

In Scenario B, the null hypothesis is that the survey unit meets the release criterion. If the measurement distribution is highly skewed, there may a substantial portion of the survey unit with residual radioactivity higher than the $DCGL_W$, but perhaps not in excess of the $DCGL_{EMC}$. In such cases, the median may be below the LBGR, while the mean is above it. The *Quantile* test, discussed in Section 2.5.5, can be used to detect when remediation activities have failed in only a few areas within a survey unit. Conducting the Quantile test in tandem with the WRS has been found to improve the accuracy of the tests (EPA 230-R-94-004, 1994).

In some cases, data from an asymmetric distribution can be transformed so that the transformed data have a more symmetric distribution. The analysis is then performed using the transformed data. A common example is that the logarithms of lognormally distributed data have a normal distribution. However, such transformations introduce additional complications. The mean of the transformed data is not generally equal to the transform of the mean of the original data. For example, the mean of the logs transforms back to the geometric mean, which is the median of a lognormal data set. The computations necessary for testing the average of lognormal data can be complex (see Chapter 14). The behavior of this lognormal test when the assumption of lognormality is violated is not known.

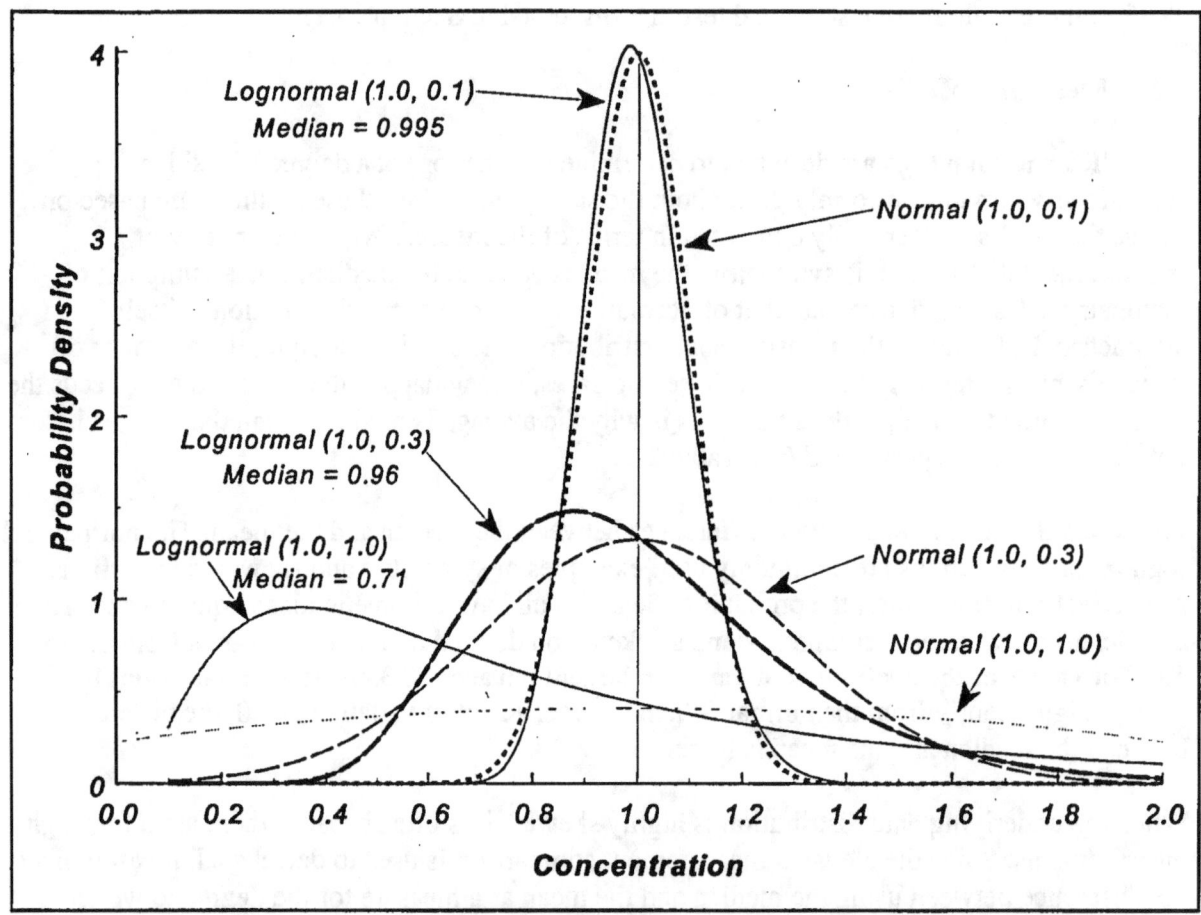

Figure 2.5 Comparison of Normal and Lognormal Distributions
(Arithmetic mean and arithmetic standard deviation shown in parentheses)

The EPA (EPA QA/G-4, 1994) has compared the use of the mean and the median as the parameter of interest whose true value the decision maker would like know and that the data will estimate. Some of the positive and negative attributes of each are listed below:

MEAN

Positive Attributes

- Useful when action level is based on long-term, average health effects
- Useful when the population is uniform with relatively small spread.
- Generally requires fewer samples than other parameters.

Negative Attributes

- Not a very representative measure of central tendency for highly skewed populations.
- Not useful when the population contains a large proportion of values that are less than measurement detection limits.

MEDIAN

Positive Attributes

- Useful when action level is based on long-term, average health effects
- More representative measure of central tendency than the mean for skewed populations.
- Useful when the population contains a large number of values that are less than measurement detection limits.
- Relies on few statistical assumptions.

Negative Attributes

- Will not protect against the effect of extreme values.
- Not a very representative measure of central tendency for highly skewed populations.

2.5.4 Estimating the Amount of Residual Radioactivity

The result of the statistical test is the decision whether or not to reject the null hypothesis. Following the statistical hypothesis tests, it may also be necessary to estimate the amount of residual radioactivity in the survey unit so that dose calculations can be made. It is usually best to use the mean (average) residual radioactivity for this purpose (EPA PB92-963373, 1992). If the data distribution is symmetric, the mean is equal to the median. If, however, the data distribution is skewed, the mean may be greater than the median.

2.5.5 Quantile Test

In contrast to locations with concentrations above the $DCGL_{EMC}$ more moderate departures from uniformity in residual radioactivity concentrations may also exist within a survey unit. One portion of a survey unit may have virtually no residual radioactivity, while another portion does contain some residual radioactivity. There may be several portions of one type or another in a survey unit, resulting in a patchy contamination pattern. The Quantile test is designed to detect this type of residual radioactivity. The Quantile test is only needed in Scenario B.

Like the WRS test, the Quantile test (EPA 230-R-94-004, 1994; Johnson et al., 1987) is a two-sample test. It is also performed by first subtracting the value of the LBGR from each of the survey unit measurements, and then listing the combined set of adjusted survey unit and reference area measurements from smallest to largest. However, only the largest measurements in the list are examined. The number of measurements that will be considered in the Quantile test is denoted by r. A count is made of the number of measurements among the largest r measurements that are from the survey unit. This number is denoted by k. If there is no contamination, measurements from the reference area and from the survey unit would be expected to appear among the r largest measurements roughly in proportion to the number of measurements made in each of them. If patchy residual contamination exists, then the r largest measurements of the combined data sets (reference area and survey unit) are more likely to come from the survey unit. When there are m background measurements and n survey site measurements, then k should be about r times $n/(m+n)$. If the number of measurements from the survey unit among the largest r is too much larger than this, then there is evidence that the survey unit contains patchy areas of residual radioactivity in excess of the LBGR.

While it is possible to perform a one-sample version of the Quantile test, it will seldom be necessary in practice. With no interfering background, patchy areas of contamination in excess of the LBGR will be apparent in simple posting plots and histograms (see Chapter 4) of the survey unit data.

2.5.6 Elevated Measurements Comparison

An *elevated measurement comparison* is performed by comparing each measurement from the survey unit to the $DCGL_{EMC}$. If the survey unit is being compared to a reference area, the net survey unit measurement is first obtained by subtracting the mean of the reference area measurements. A net survey unit measurement that equals or exceeds the $DCGL_{EMC}$ is an indication that a survey unit may contain residual radioactivity in excess of the release criterion.

This type of measurement comparison is sometimes called a "hot spot test." The latter term may be misleading because it is not a formal statistical test, but a simple comparison of measured values against a limit. In addition, there is no commonly accepted definition of what constitutes a *hot spot* in either area or magnitude of residual radioactivity. Yet, this term may imply some degree of radiological hazard. In this report, the term "area of elevated residual radioactivity" is used to describe a limited area of residual activity that may cause the decommissioning dose criteria to be exceeded. It is only these areas that might be considered *hot spots*. For planning purposes, the potential extent of an "area of elevated residual radioactivity" is based on the distance between sampling points in the survey sampling grid.

In addition to direct measurements or samples at discrete locations, parts of each survey unit will also be scanned. For the quantitative measurements obtained at discrete locations, performing the EMC is a straightforward comparison of two numerical values. Some sophisticated scanning instrumentation is also capable providing quantitative results with a quality approaching those from direct measurements or samples. Other scanning measurements, however, may be more qualitative. In that case, *action levels* should be established for the scanning procedure so that areas with concentrations that may exceed the $DCGL_{EMC}$ are marked for a quantitative measurement.

A single unusually large measurement may occur by chance. The elevated measurement comparison flags these measurements for further study. When a measurement is flagged, it should first be determined that it is not due to sampling or analysis error. Such a determination may include re-sampling the area at which the measurement was originally taken.

If an measurement exceeding the is $DCGL_{EMC}$ is confirmed, then the size of the area of elevated residual radioactivity, A', and the average concentration within it, $C_{A'}$, is determined. This will generally involve taking further measurements in the vicinity of the elevated measurement to adequately delimit its extent. Using the area factor F_A, for the area A', C_A, should not exceed the product $(F_A)(DCGL_W)$ in order for the survey unit to meet the release criterion.

2.5.7 Investigation Levels

In contrast to an elevated measurement, a measurement may be found that exceeds the concentration level expected from the survey unit's classification. *Investigation levels* are established for each class of survey unit to guard against the possible mis-classification of survey units. If a measurement exceeds the investigation level, additional investigation is required to determine if the final status survey for the survey unit was adequate to determine compliance with the release criteria.

For example, in a Class 1 survey unit, measurements above the $DCGL_W$ may not be unusual or unexpected. In Class 2 areas, however, neither measurements above the $DCGL_W$ nor elevated areas are expected. Thus in these areas, any measurement at a discrete location exceeding the $DCGL_W$ should be flagged for further investigation. Unless the scanning sensitivity is such that an action level can be specified for areas with concentrations potentially exceeding the $DCGL_W$, any positive indication of residual radioactivity during the scan could warrant further investigation.

Because there is a low expectation of any residual radioactivity in a Class 3 area, it may be prudent to investigate any measurement exceeding even a fraction of the $DCGL_W$. What level that should be will depend on the site, the radionuclides of concern, and the measurement and scanning methods chosen. This level should be set using the DQO Process during the survey design phase of the Data Life Cycle. In some cases it may be prudent to follow this procedure for Class 2 survey units as well.

Suggested investigation levels that might be appropriate for each class of survey unit and type of measurement are shown in Table 2.4.

Table 2.4 Summary of Investigation Levels

Survey Unit Classification	Flag Direct Measurement or Sample	Scanned Area Marked When Action Levels Indicate:
Class 1	$> DCGL_{EMC}$	$> DCGL_{EMC}$
Class 2	$> DCGL_W$	$> DCGL_W$
Class 3	$>$ fraction of $DCGL_W$	$> MDC$

In the last two sections, we have considered elevated measurements that require investigation for compliance with the dose criteria, and investigation levels that flag potential survey unit mis-classifications. In addition to these, are the QA/QC procedures that should be in place in any measurement program. Gross errors of analysis can often be spotted by the use of simple preliminary data analysis techniques, such as posting plots and histograms. This is sometimes called exploratory data analysis. These techniques also form a part of the Data Quality Assessment process (EPA QA/G-9, 1995). Some of these methods are discussed in Chapter 4.

3 PLANNING FINAL STATUS SURVEYS: DATA QUALITY OBJECTIVES

An essential consideration in designing survey plans for site decommissioning is that the radiological data that are collected and analyzed are sufficient and of adequate quality for decision-making purposes. *It is imperative that the type and quality of radiological data that will be needed to support license termination be considered early in the decommissioning process.*

Before commencement of survey work, it is essential that a survey plan be developed that is based on the data needed for decision making and the level of quality needed to support the decision. Such a plan should specify what samples need to be obtained, how and where they will be collected and analyzed, what quality assurance procedures will be used, the method of comparing site areas to reference areas, and what level of decision errors will be considered acceptable. These decisions become paramount for determining compliance with decommissioning criteria that are near background levels.

3.1 Introduction

The Data Quality Objectives (DQO) process is a series of planning steps based on the scientific method that is designed to ensure that the type, quantity, and quality of environmental data used in decision making are appropriate for the intended application (EPA QA/G-4, 1994). DQOs are qualitative and quantitative statements that

- clarify the study objective
- define the most appropriate data to collect
- determine the most appropriate conditions for collecting the data and
- specify acceptable levels of decision errors that will be used as the basis for establishing the quantity and quality of data needed to support the decision.

The DQO process comprises the following steps:

(1) State the problem, i.e., the objective of the sampling effort.
(2) Identify the decision, i.e., the decision to be made that requires new data
(3) Identify inputs to the decision, i.e., the data that are needed and how they will be used to support the decision.
(4) Define the study boundaries, i.e., the spatial and temporal aspects of the environmental media that the data represent.
(5) Develop a decision rule, i.e., an "if...then" statement that defines the conditions for choice among alternative actions.
(6) Specify limits on decision errors.
(7) Optimize the design for obtaining data, i.e., the most time- and resource-effective sampling and analysis plan.

The DQO process is iterative, so that any and all of the specifications may change as new information is obtained during the course of site remediation, up until the final status survey is actually performed.

It is important to specify the type and quality of radiological data that will be needed for final status surveys *early* in the decommissioning process. This process entails early specification of sample collection and analysis procedures, the determination of DCGLs, the classification of survey units, the method of comparing survey units to reference areas, the null and alternative hypotheses, Type I and Type II error rates, and quality assurance procedures.

In the following sections, each of the seven steps in the DQO process is discussed as it pertains to the planning, design, and performance of the final status survey.

3.2 State the Problem

For most NRC licensees, the objective of the decommissioning process is to remove their facilities safely from service and reduce residual radioactivity to a level that permits release of the property and termination of the license. The data that will be needed to support this objective will demonstrate that any residual radioactivity remaining on the site results in a dose that does not exceed the release criterion. *This objective will be met by performing a final status survey in individual survey units. For each survey unit, a separate decision will be made on the attainment of the release criterion.*

The final status survey occurs near the end of the decommissioning process, following historical site assessment, scoping, characterization, and remediation. These earlier steps in the decommissioning process provide crucial information for the design of the final status surveys. This information includes the identification of potential residual radioactive materials, the general locations and extent of residual radioactivity, and estimates of the concentration levels and its variability. Some of this information may be part of the licensee's decommissioning plan.

3.3 Identify the Decision

For the final status survey, the essential decision is whether the decommissioning criteria have been met in individual survey units. The decommissioning criterion is expressed in terms of a total effective dose equivalent (TEDE) limit above background due to residual radioactivity. The decision will be based on radiological data collected in a survey designed for this purpose. Procedures for the design of the final status survey and for the statistical analysis of the results are the primary focus of this report.

An essential part of identifying the decision, is a knowledge of the applicable residual radioactivity concentration limits. These are the Derived Concentration Guideline Levels (DCGLs) discussed in Section 2.2.1. NRC has developed models to provide generic dose conversion factors for residual radioactivity that can be applied within a hierarchy of modeling approaches. The models provide a mechanism for translating the residual radioactivity at a site into TEDE using the site-specific source term and varying levels of related site information. The provisions of 10 CFR Part 20, Subpart E require that a licensee consider the entire applicable source term and all credible dose pathways when determining whether any residual radioactivity meets the decommissioning criteria.

Since the dosimetric models are used to define the TEDE release criterion in terms of a DCGL, careful consideration should be to the assumptions made in those models. Screening models are generally the easiest to use. These models are constructed to cover a wide range of possible conditions, and so are also generally the most conservative. Their use may result in very low DCGLs. Using site specific parameters in such a model can reduce this conservatism considerably, but will require some justification. A balance should be sought between the complexity of site specific modeling and the potential cost in remediation and surveys of using DCGLs that are overly conservative.

For sites which contain residual radioactivity distinguishable from background from more than one radionuclide, there are two methods that can be used. If the concentrations of the radionuclides within a survey unit are related, one radionuclide can be used as a surrogate for the others using a modified $DCGL_W$. Otherwise, the TEDE due to the mixture of radionuclides is compared to the release criterion by applying a *mixture rule*. This is done by determining the ratio between the concentration of each radionuclide in the mixture and the $DCGL_W$ for that radionuclide alone. The sum of the ratios for all radionuclides in the mixture should not exceed one. The case of multiple radionuclides is discussed further in Chapter 11.

3.4 Identify Inputs to the Decision

Although the final status survey is performed near the end of the decommissioning process, it is possible to produce a more efficient survey design if the requirements of this survey are identified early in the decommissioning planning. By knowing in advance the type, quantity, and quality of data that are needed in the final status survey, information obtained from earlier decommissioning surveys may be used to support the final status survey.

Previous steps in the DQO process have identified the critical radionuclides, and established their corresponding concentration or surface activity limits (DCGLs) for various post-remediation land use scenarios. In subsequent steps, acceptable limits on decision errors, and the number of measurements necessary to meet them, will be established. To accomplish this, an estimate of the expected variability of the measurement data will be needed. Information from scoping, characterization, and remediation control surveys can be very useful for estimating the mean and standard deviation expected for residual radioactivity in a survey unit and for background radioactivity in one or more reference areas. In the absence of such data, experience and scientific judgment can be used to estimate the expected measurement variability or a separate scoping survey may be conducted. The effort required for an adequate estimate of the expected measurement variability will depend on its magnitude relative to the DCGLs. The smaller the value of the DCGL relative to the expected measurement standard deviation, the more important it will be to have an accurate estimate of that standard deviation. Thus, surveys performed earlier in the decommissioning process can provide valuable information for designing the final status survey. As more information comes available, both the measurement and statistical methods that will be needed to meet release criteria can be refined.

The selection and proper use of appropriate instruments and techniques will be critical factors in assuring that the survey accurately determines the radiological status of the site. In this report, three basic types of measurements are considered:

(1)   scanning
(2)   direct field measurements
(3)   laboratory analysis of samples.

Scanning is the process by which the surveyor moves a portable radiation detection instrument over a surface (i.e., ground, wall, floor, equipment) to detect the presence of radiation. A scan is performed to locate radiation anomalies that might indicate elevated areas of residual activity that will require further investigation or action. If scan survey results exceed a scanning action level determined on the basis of the potential contaminant and the detector and survey parameters, the location is noted for further action (direct measurement or sampling).

Direct field measurements are those made at a fixed location using portable instruments (e.g., survey meter, pressurized ionization chamber (PIC), *in situ* spectrometer). The result of a direct measurement, as opposed to a scan, is a quantitative measure of the radioactivity present at the location measured.

Taking samples, with subsequent analyses conducted in a laboratory, will be required for certain radionuclides and radiations that cannot be adequately detected using direct measurements. For some nuclides or environmental media, this may be the only realistic technique to employ.

The analysis techniques used may be *radionuclide specific* or for *total radioactivity*.

The survey designs with which these measurements are made fall into two categories:

(1)   authoritative (judgment) sampling
(2)   probability sampling

Authoritative or *judgment sampling* occurs when measurements are made or samples are collected at locations where anomalous radiation levels are observed or suspected. The term "biased sampling" is sometimes used to indicate that the sample locations are not chosen on a random or systematic basis. Biased radiological measurements and samples also may be taken to further define the areal extent of potential contamination and to determine maximum radiation levels within an area.

When data quality objectives involve statistical estimation or hypothesis testing, some form of *probability sampling* is required. The type of probability sampling recommended for use in final status surveys is either simple random sampling (for Class 3) or systematic sampling on a systematic grid with a random start (for Class 1 and Class 2).

Of the three measurement types, only the results of direct measurements and sampling are used in conducting the nonparametric statistical tests. All three types of measurement result are subject to an elevated measurement comparison against an upper limit value.

The type of instrumentation or sampling and analysis methodologies or both used for final status surveys will influence the number of samples or direct measurements, or both, that are required for the appropriate statistical analysis of the data. As a rule, the less precise the measurement, the greater the number of measurements that will be required for the statistical tests to achieve the desired level of uncertainty. The selection of survey instruments may involve a cost analysis of whether it is better to use a more precise (and more expensive) measurement method with correspondingly fewer measurements, or to use a less precise (and perhaps less costly) method that would require the collection of more measurements. The information necessary to calculate the required number of samples, given the expected variability of the data, is discussed in discussed in Section 3.7.

Similar considerations are involved in the choice of making radionuclide-specific measurements versus total alpha, beta, or gamma activity or total exposure rate measurements or both. If total (gross) methods are used, the results will include the variability of natural background. This additional variability will not only require more measurements to overcome but will also necessitate comparison with a reference area using the two-sample Wilcoxon Rank Sum test of Chapter 6 rather than the one-sample Sign test of Chapter 5.

If the radionuclide of concern appears as part of background, there is no alternative to a survey unit comparison to a reference area; however, the measurement precision will still affect the number of samples required. Radionuclide-specific methods should be considered in this case as well, since the variability of the total activity present will be greater than that due to any particular radionuclide or series alone.

Instrumentation can be selected using guidelines that compare its performance capabilities to the applicable decommissioning criteria. Consideration should be given to the characteristics of the type of detector, in particular, the minimum detectable concentration (MDC) for the radionuclide under investigation. The simplest of devices, survey meters, may be appropriate for hand scanning of building surfaces for certain nuclides at certain activity levels. Fixed-place detectors at grid points can be used in other situations. In some situations, the sensitivity needed at background levels will require that measurements be nuclide specific, thereby requiring spectrometric techniques. Consideration should also be given to newer technologies as they are developed.

3.5 Define the Study Boundaries

Defining the spatial and temporal boundaries will help ensure that the samples taken in the survey are representative of the survey unit for which the decommissioning decision will be made. Spatial boundaries describe what measurements or samples should be taken and in what areas. Temporal boundaries describe when the measurements or samples should be taken, and any time constraints on the data collection and analysis. Uniformity over a given area should be checked wherever possible. This can be done by inspecting the site and knowing its history from data collected earlier in the decommissioning process, or by scanning measurements. The selection of measurement and sampling points must ensure that the sample is representative of the site category under investigation.

3.5.1 Spatial Variability

As has been discussed in Sections 2.2.6 and 2.2.7, some estimate of the variability of the data is needed for a good survey design. The smaller the variability within each reference area or survey unit, the smaller the number of samples that will be needed to achieve the specified Type I and Type II error rates for the test. Thus, it is advantageous to identify survey units that are relatively homogeneous in radiological character. Reference areas and survey units should be as similar as possible with regard to their background characteristics.

Considering the variability in collected data that is expected in any environmental sampling program, accurate interpretation of the results is essential. The choice of individual survey units and any reference areas to which they are to be compared is especially significant. In the analysis of the data, any systematic difference in the measurements from a survey unit and a referenece area is assumed to be due to residual radioactivity. The choice of a reference area is a spatial extrapolation of the background radionuclide concentrations there to the survey unit. It would be obviously inappropriate to compare uranium concentrations in soils collected from two sites of different geology, such as a sandy beach area and an inland region with heavy clay soil. In the case of the fallout radionuclide ^{137}Cs, concentrations in surface soils could only be extrapolated to other local plots of land that have received the same deposition (rainfall) and have the same history (for example, plowed agricultural land, forest, or undisturbed lawn). For instance, the presence of ^{137}Cs in soil, and the observation that it is not at the same level from place to place, does not necessarily indicate a local facility contribution. Such variations may have resulted from disturbance to the site through either natural or human action, which led to removal or addition of material containing fallout from atmospheric nuclear weapons tests, as well as differences in the spatial distribution of the original deposition.

In some situations involving radionuclides that appear as part of natural background, the screening level DCGLs may be small compared to the spatial variations among even nearby and closely matched reference areas and survey units. In such cases, an effort should be made to reduce exposure pathway modeling conservatism by using site specific parameters and realistic occupancy scenarios. In particularly difficult cases it may be necessary to explore alternative statistical methods for establishing whether residual radioactivity in a survey unit is distinguishable from background. An example of such an analysis is given in Chapter 13.

3.5.2 Temporal Variability

Temporal variability will contribute to the overall uncertainty of comparisons of survey units and reference areas, although generally to a lesser extent than spatial variability. However, it is best to avoid temporal variability to whatever extent possible. This might be accomplished by collecting data from areas to be compared over as short a time interval as possible, and avoiding circumstances known to cause short-term background variations. There may be reasons why samples cannot be taken in certain places or at certain times. These constraints should be identified so that they can be accounted for in the planning process.

3.5.3 Reference Coordinates

As part of this step in the DQO process, a site diagram should be prepared showing each potential survey unit and the reference area to which it will be compared. For each unit, the types of samples that will be taken, the analyses needed, and a schedule for sampling and analysis should be listed.

Reference coordinate systems are established at the site to facilitate selection of measurement/sampling locations and provide a mechanism for referencing a measurement to a specific location so that the same survey point can be relocated.

A survey reference coordinate system consists of a grid of intersecting lines, referenced to a fixed site location or benchmark. Typically, the lines are arranged in a perpendicular pattern, dividing the survey location into squares or blocks of equal area. Reference coordinate system patterns on horizontal surfaces may be identified numerically on one axis and alphabetically on the other axis or in distances in different compass directions from the grid origin. Interior walls are treated as extensions of the floor along the horizontal plane. An example of a structure interior reference coordinate system using letters and numbers in shown in Figure 3.1. An example land area reference coordinate system using distance from an origin along north-south and east-west lines is shown in Figure 3.2.

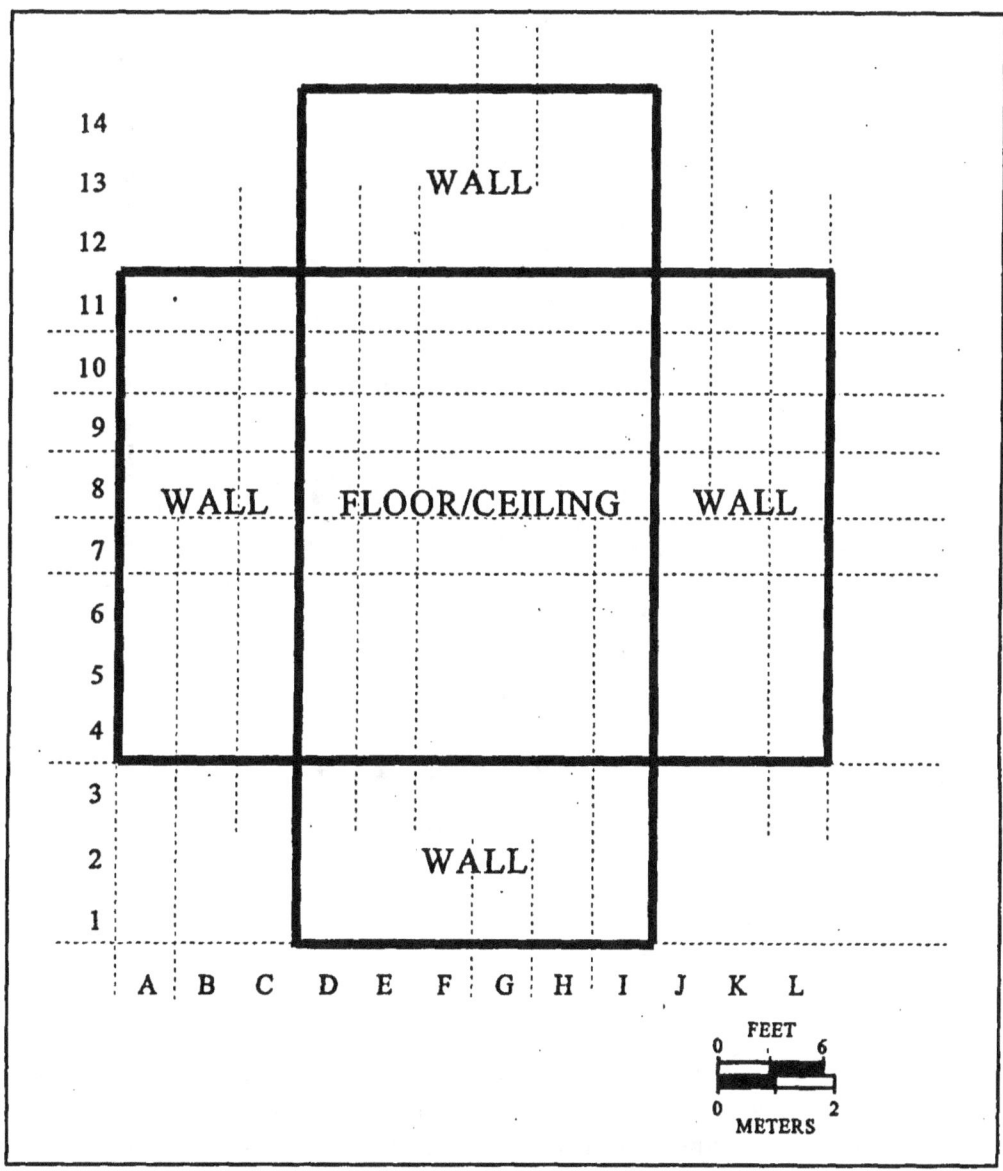

Figure 3.1 Sample Indoor Reference Coordinate System.

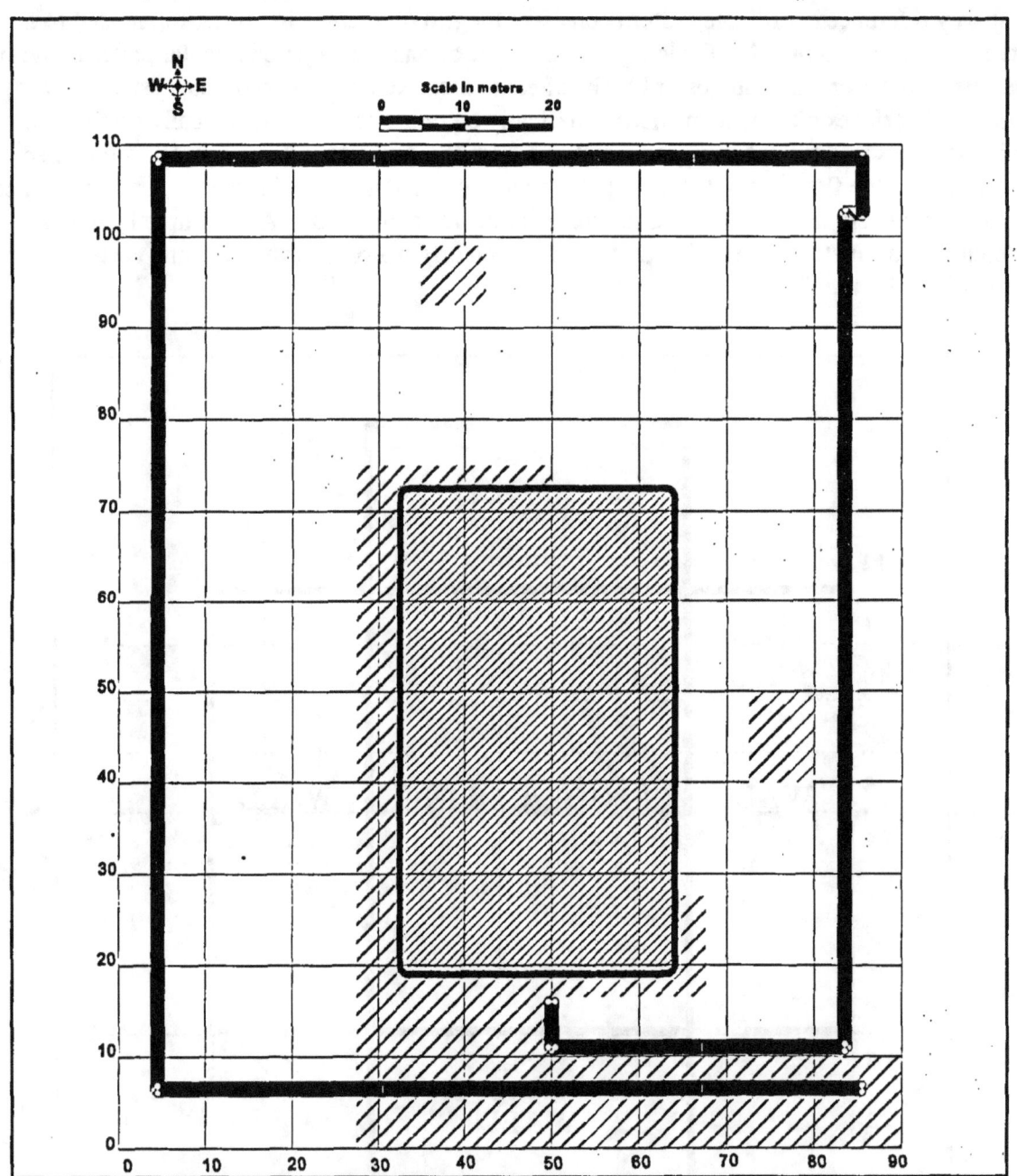

Figure 3.2 Sample Outdoor Reference Coordinate System

One procedure for constructing a reference coordinate system is to draw a map of the area to be sampled and a rectangle that encloses it. Define a coordinate system for locating points (X,Y) within the rectangle, e.g., the number of meters east, X, and the number of meters north, Y, from the southwest corner (0,0) of the rectangle. The northeast corner will then have coordinates (X_{max}, Y_{max}). Note that the local coordinate system need not line up with the principal compass points. It may be convenient to align one of the axes with a site boundary or other local feature.

For the example in Figure 3.2 the coordinate system has been laid out in the north-south and east-west directions. There are 9 ten-meter east-west coordinates, and 11 ten-meter north-south coordinates. The total area is 9,900 m², of which approximately 9,000 m² is the affected area within the fence line. The soil area to be surveyed is about 4,500 m². The remainder of the area is covered by buildings, walkways, etc., which will be part of other survey units.

3.5.4 Sampling Grids

Sampling locations in Class 1 and Class 2 survey units are laid out on random start systematic grids. The essential procedure for determining where samples should be taken in either reference areas or survey units is the same. On a site map, a reference coordinate system is laid out as in Section 3.5.3, with enough detail to locate positions with an error that will be small compared to the distance between samples. A square or equilateral triangular systematic sampling grid pattern is superimposed on the coordinate system. The length, L, of a side of either the square or the triangle used to generate the pattern, is the distance between sampling locations. This distance is determined by the total number of samples or measurements to be taken. This number, n, is calculated to satisfy the requirements of the statistical tests and is discussed in Section 3.8.1. The length (or spacing), L, of the systematic pattern is given by:

$$L = \sqrt{\frac{A}{0.866\, n}} \quad \text{for a triangular grid,}$$

and

$$L = \sqrt{\frac{A}{n}} \quad \text{for a square grid}$$

where A is the area of the survey unit.

After L is determined, a random coordinate location is identified as the starting location for the survey pattern. Beginning at the random starting coordinate, a row of points is identified, parallel to the X axis, at intervals of L.

For a *square* grid, the second and subsequent rows of points are spaced at intervals of length L along the Y axis.

For a *triangular* grid, the second row of points is located parallel to the first row, but at a Y-axis distance of $0.866L$ from the first row. The survey points along the second row are located midway (on the X-axis) between the points on the first row. This process is then repeated to fill out the triangular pattern across the survey unit.

If points are identified that fall outside the survey unit or at locations which cannot be surveyed, additional points are determined, using the same random process as was used to determine the starting point, until the desired total number of points is identified.

An example based on Figure 3.2 is shown in Figure 3.3. The procedure used for a laying out the triangular sampling grid for the soil area survey unit is as follows:

(1) Locate a random starting point by drawing two random numbers from a uniform distribution on the interval [0,1]. Random numbers can be generated using the random number function of a spreadsheet or a scientific calculator. Table A.6 contains 1000 random numbers generated using a spreadsheet, and similar tables can be found in many statistics texts. Choose any starting point in the table, and then take numbers consecutively either across rows or down columns. For example, in Table A.6, starting at row 23 in column 2 and working down, the two numbers 0.93062 and 0.029842 are found. Scale the first number by the length of the east-west coordinate axis to get 83.76 = (90)(0.93062). Round the coordinates to the nearest values that can be easily measured in the field (e.g., nearest meter). This gives 84 meters to the nearest meter. Similarly scale the second number by the length of the north-south coordinate axis to get 3.28 = (110)(0.029842) or 3 meters to the nearest meter. This gives (84,3) as the starting coordinate for the sampling grid. Since this does not fall within the area to be sampled (it falls on an area of asphalt), the next two random numbers (0.863244,0.921291) are taken, giving (78, 101). Continue until a point that falls within the sampling area is obtained. In this case (78, 101) does fall in the area to be sampled. The points are shown on Figure 3.3.

(2) Compute the spacing, L, of the sampling locations on the triangular grid using the number of sampling locations required, n, *rounded down* to the nearest meter. Rounding down helps assure that the requisite number of sampling points are identified on the sampling grid.

$$L = \sqrt{\frac{A}{0.866n}} = \sqrt{\frac{4500}{(0.866)(17)}} = 17.5 \text{ meters} \approx 17 \text{ meters}$$

Note that the area, A, is the net area of the survey unit, i.e. with buildings and paved areas that are not part of the soil area survey unit subtracted.

(3) From the starting location, lay out a row of sampling points parallel to the X-axis and distance L apart, as is shown in Figure 3.3.

(4) To start additional rows, locate the midpoint between two adjacent sampling locations on the sample row and mark a spot at a distance

$$0.866\sqrt{\frac{A}{0.866n}} = \sqrt{\frac{(0.866)(4500)}{(17)}} = 15.14 \text{ meters} \approx 15 \text{ meters}$$

perpendicular to the row. Again, this number should be rounded *down* if necessary. This is the starting location for the new row. This is also shown in Figure 3.3.

(5) Continue until all grid points within the sampling area have been located. Ignore any sampling locations that fall outside the area to be sampled. The completed sampling grid is shown in Figure 3.4.

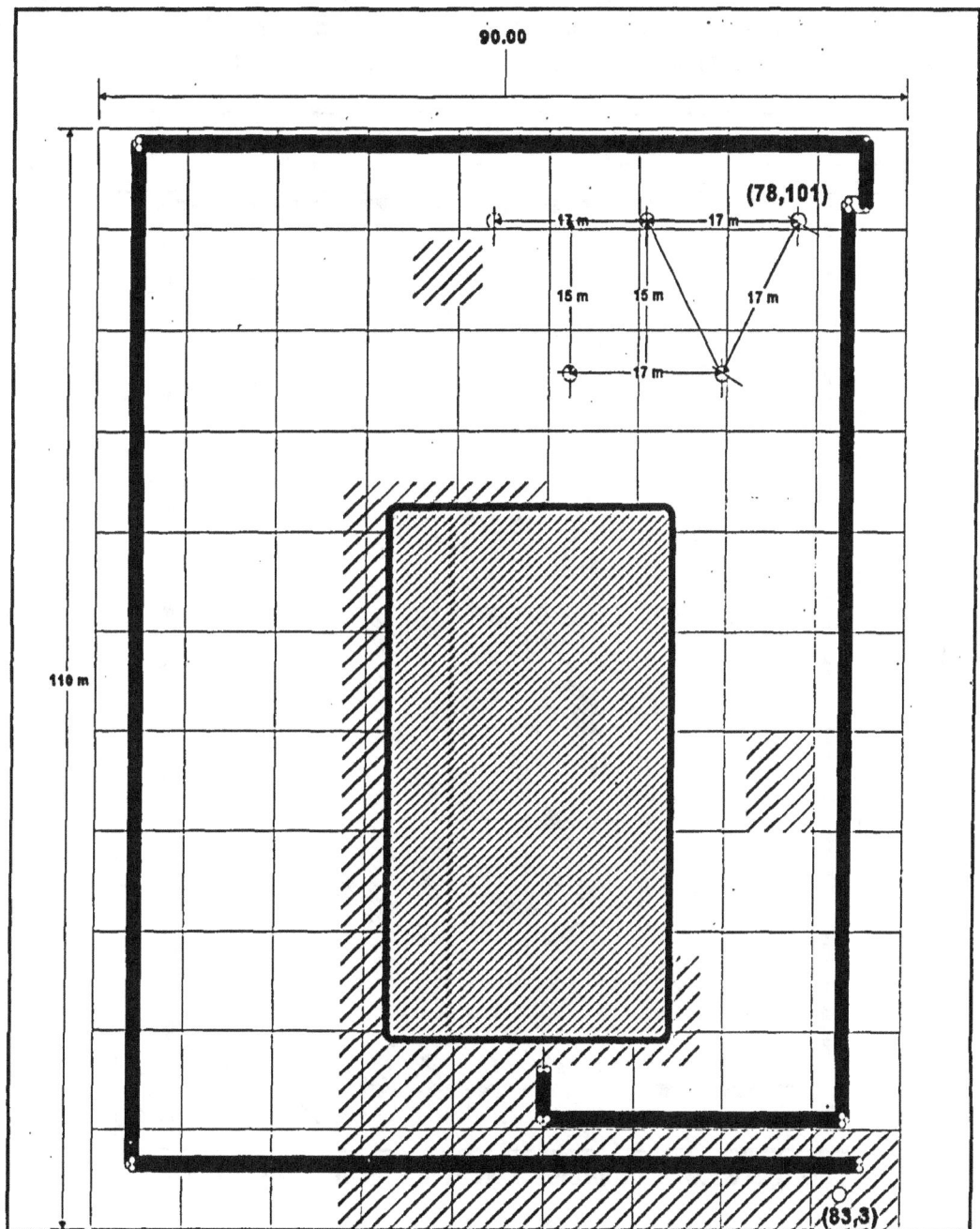

Figure 3.3 Laying Out a Triangular Grid

Using this procedure, the number of sampling points on the triangular grid within the sampling area may differ from the desired number, *n,* depending on the shape of the area. In this example, because of the irregular shape of the region caused by its wrapping around the building, 20 sampling points are found on the grid. If the number of points is greater than the desired number, use all the points.

If the number of points is less than the desired number, the additional points required may be identified using the same procedure that was used to determine the grid starting point. These will be at individual random locations within the sampling area, and should be used regardless of where they occur relative to the grid. If more than a few random sample locations are needed, it is preferable to lower the grid spacing, L, and redraw the sampling grid.

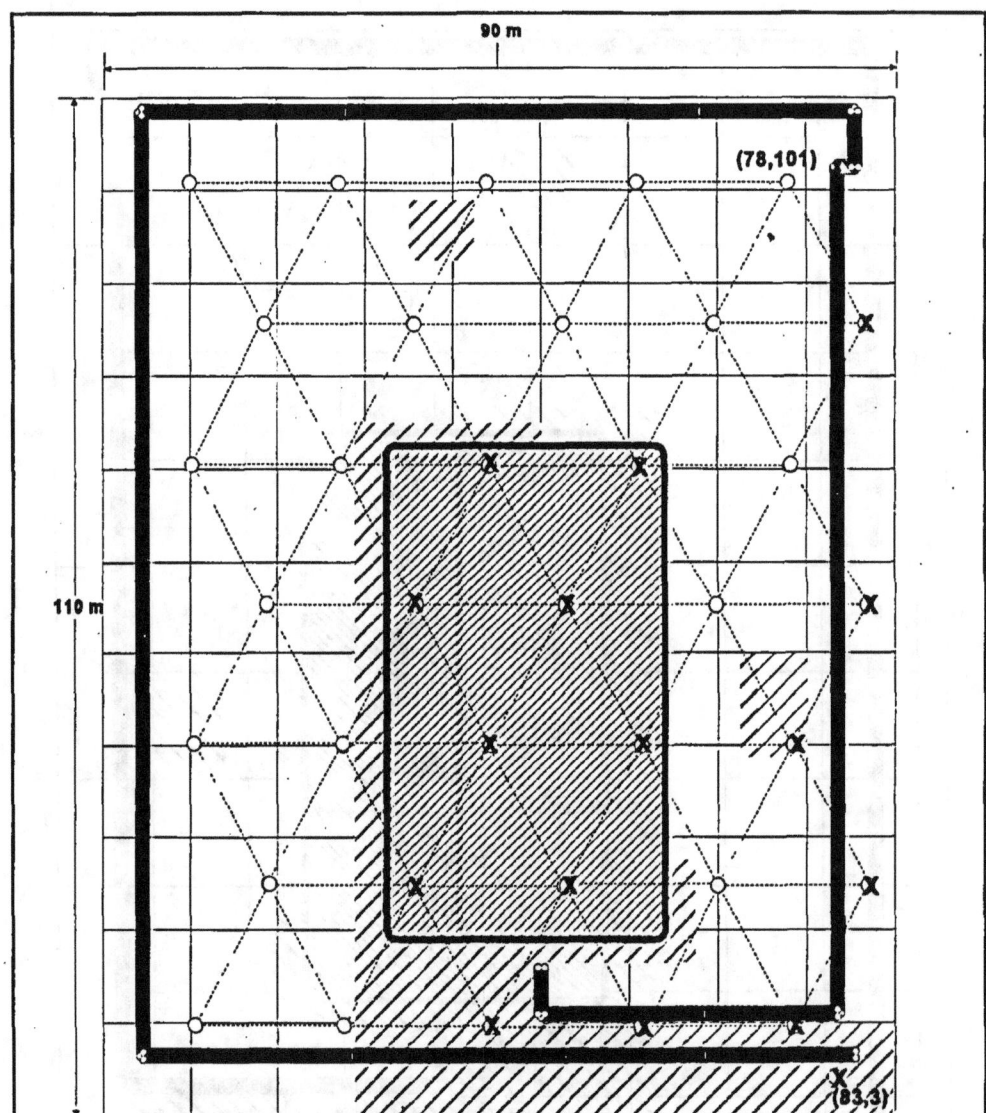

Figure 3.4   Completed Triangular Sampling Pattern

3.6 Develop a Decision Rule

A decision rule relates the concentration of residual radioactivity in the survey unit to the release criterion so that decisions can be made based on the results of the final status survey.

The decision rule proposed in this report for the final status survey consists of a statistical test and an elevated measurement comparison. The specific recommended statistical tests were discussed in Section 2.4, and the elevated measurement comparison was discussed in Section 2.5. Alternative statistical tests may be appropriate in specific circumstances, providing the assumptions of those tests are verified.

In some cases, it will not be necessary to formally conduct the statistical tests. If the radionuclide is not in background and radionuclide-specific measurements made, the survey unit meets the release criterion if all of the measurements are below the Derived Concentration Guideline Level for the mean residual radioactivity ($DCGL_w$) defined in Section 2.2.1. On the other hand, if the average of the measurements is above the $DCGL_w$, the survey unit does not meet the release criterion. It is only when some measurements are above the $DCGL_w$, but the average is below the $DCGL_w$, that the Sign test and the elevated measurement comparison need to be used.

If the radionuclide appears in background or if non-radionuclide-specific measurements made, the survey unit always meets the release criterion when the difference between the maximum survey unit measurement and the minimum reference area measurement is below $DCGL_w$. If the difference between the survey unit and reference area averages is above $DCGL_w$ the survey unit fails to meet the release criterion. When the maximum difference is above the $DCGL_w$ but the average difference below $DCGL_w$, conduct the WRS test and elevated measurement comparison.

Recall that the $DCGL_w$ is the concentration level corresponding to the release criterion when the residual radioactivity is spread throughout the survey unit rather than is smaller elevated areas. The Lower Boundary of the Gray Region (LBGR) is the concentration level below which further remediation is not reasonably achievable. The null and alternative hypotheses for the statistical tests that were discussed in Section 2.3.1 can be restated in terms of the $DCGL_w$ and LBGR as follows.

Scenario A
Null Hypothesis:
H_0: The mean concentration of residual radioactivity above background in the survey unit exceeds the $DCGL_w$.
versus
Alternative Hypothesis:
H_a: The mean concentration of residual radioactivity above background in the survey unit does not exceed the LBGR.

Scenario B
Null Hypothesis:
H_0: The mean concentration of residual radioactivity in the survey unit is indistinguishable from background up to a level specified by the LBGR.
versus
Alternative Hypothesis:
H_a: The mean concentration of residual radioactivity in the survey unit distinguishable from background is in excess of the $DCGL_w$.

These hypotheses are stated in terms of the mean concentration, which is the parameter of interest. As discussed in Section 2.5, nonparametric tests are used to test these hypotheses, using the elevated measurement comparison to correct potential inaccuracies when the measurement distribution is very skewed.

The preceding paragraphs have indicated some decision rules for final status surveys. However, there has been as yet no statement about how confident one can be that the decision is correct. While a formal statistical test may, in retrospect, not be needed to analyze the data, the survey must always be designed as if it were. Otherwise there is no basis for deciding the number of measurements to be taken, or with what precision. Nor will there be any basis for confidence in the resulting decision. This is the subject of the next Section. The hypothesis testing framework allows a estimate to be made of the Type I and Type II error probabilities. In addition, it is possible to similarly calculate the probability that the null hypothesis will be rejected (i.e., the power of the test) at any specific residual radioactivity concentration level, given some assumptions concerning the distribution of the residual radioactivity. In Section 3.7 on specifying the limits on decision errors, it will be seen that this allows the ALARA concept to be explicity incorporated into the decision-making process.

A different sort of decision rule is required for the elevated measurement comparison. As indicated in Section 2.2.1, the value of the $DCGL_{EMC}$ is based on a specific size area of elevated residual radioactivity. The area used during the survey planning to determine the $DCGL_{EMC}$ is based on the distance between the sampling locations on a systematic grid. The *actual* extent of an elevated area cannot be determined from a single measurement. When a measurement exceeds the $DCGL_{EMC}$, further investigation is required to determine both the size of the elevated area and its average concentration of residual radioactivity. Only then can it be determined if the TEDE due to the elevated area exceeds the release criterion. Thus, the decision rule for the EMC is a two stage process. In the first stage, areas are flagged as potentially elevated at specified investigation levels. In the second stage, the actual average concentration over the actual extent of elevated area is compared to the release criterion. The level at which measurements should be flagged depends on the survey unit classification, as discussed in Section 2.5.7.

3.7 Specify Limits on Decision Errors

A statistical decision error occurs when the null hypothesis is rejected when it is true (Type I), or not rejected when it is false (Type II). The source of the uncertainty leading to such errors is the measurement variability, σ, discussed in Section 2.6. While the possibility of a decision error can never be totally eliminated, it can be controlled. Limits on decision errors are set to establish performance goals for the survey design. The two types of decision errors are classified as Type I and Type II decision errors, and were summarized in Table 2.2 for Scenario A and Scenario B. The probability of making a Type I decision error, or the level of significance, is called alpha (α). The probability of making a Type II decision error is called beta (β). The *power* of a test ($1-\beta$) is the probability of rejecting the null hypothesis when it is false.

This step in the DQO process is crucial. It is at this point that the limits on the decision errors rates are developed in order to establish appropriate goals for limiting uncertainty in the data. This is done by establishing the goals for the Type I error rate and the Type II error rate. The procedure for doing this follows.

3.7.1 Type I and Type II Decision Errors for Statistical Tests

A Type I error is made when the null hypothesis is rejected when it is true. A Type II error is made when the null hypothesis is not rejected when it is false. Thus the Type I and Type II errors have different meanings about meeting the release criteria, depending on whether Scenario A or

Scenario B is being considered. In an effort to avoid confusion, it will be convenient to say that a survey unit *passes* the statistical test if it is deemed to meet the release criterion as a result of that test. Otherwise, the survey unit will be said to *fail* the statistical test. The error rates can then be expressed as the probability that a survey unit passes when it should fail or fails when it should pass. This is summarized in Table 3.1. In Scenario A, the probability that a survey unit passes when it should fail is α and the probability it fails when it should pass is β. In Scenario B, the probability that a survey unit passes when it should fail is β and the probability it fails when it should pass is α. In Scenario A and B, the roles of α and β are reversed because the null and alternative hypotheses are reversed.

Table 3.1 Summary of Types of Decision Errors

If Survey Unit passes the statistical test...	...when the True Condition of Survey Unit is...	
	That It Does Not Meet Release Criterion	That It Meets Release Criterion
under Scenario A	Type I Error (Probability = α)	Correct Decision (Power = $1-\beta$)
under Scenario B	Type II Error (Probability = β)	Correct Decision (Probability = $1-\alpha$)

If Survey Unit fails the statistical test...	...when the True Condition of Survey Unit is...	
	That It Does Not Meet Release Criterion	That It Meets Release Criterion
under Scenario A	Correct Decision (Probability = $1-\alpha$)	Type II Error (Probability = β)
under Scenario B	Correct Decision (Power = $1-\beta$)	Type I Error (Probability = α)

Acceptable error rates can be established for either scenario by determining the desired probability for passing the survey unit as a function of the concentration of residual radioactivity actually remaining in the survey unit. Using the statistical tests recommended in this report, the probability that a survey unit passes decreases as the residual radiation concentration increases. At concentrations above background near the $DCGL_w$, this probability should be low in order to be adequately protective of public health. The probability that the survey unit passes should be high whenever the concentrations are near background in order to avoid unnecessary remediation costs. Somewhere in the range between residual radioactivity concentrations of zero and the $DCGL_w$ there is often is a concentration level such that remediation below this level is not considered to be reasonably achievable. considered unreasonable. The concentration range between this lower level and the $DCGL_w$, defines a *gray region* of residual radioactivity concentrations in which the consequences of decision errors are relatively minor. The specification of a gray region is important because variability in the data may be such that a decision may be "too close to call" when the true but unknown value of the residual radioactivity concentration is very near the DCGL$_W$. The *Lower Boundary of the Gray Region,* the LBGR, is, by definition, the concentration value at which the acceptable probability of failing a survey unit when it should pass (β in Scenario A, α in Scenario B) is specified.

Specifying acceptable error rates is a means for bringing considerations of practicality directly into the decision making process. The probability limits assigned to points above and below the gray region should reflect the risks involved in making decision errors. These probabilities can then be converted to acceptable Type I and Type II error rates, α and β, using Table 3.1.

Figure 3.5 illustrates this process. For example, suppose it is considered that remediation to concentrations below one-half the DCGL$_W$ cannot be achieved with reasonable effort. The desired probability that the survey unit passes should then be set at a high value when the true residual radioactivity concentrations are at or below that level. This is the LBGR for this example. When the true concentration is at the DCGL$_W$, a small probability for passing the survey unit is desired. The line segments connecting the LBGR and DCGL$_W$ points with concentration values both higher and lower reflect the fact that the probability that the survey unit passes decreases with increasing concentration.

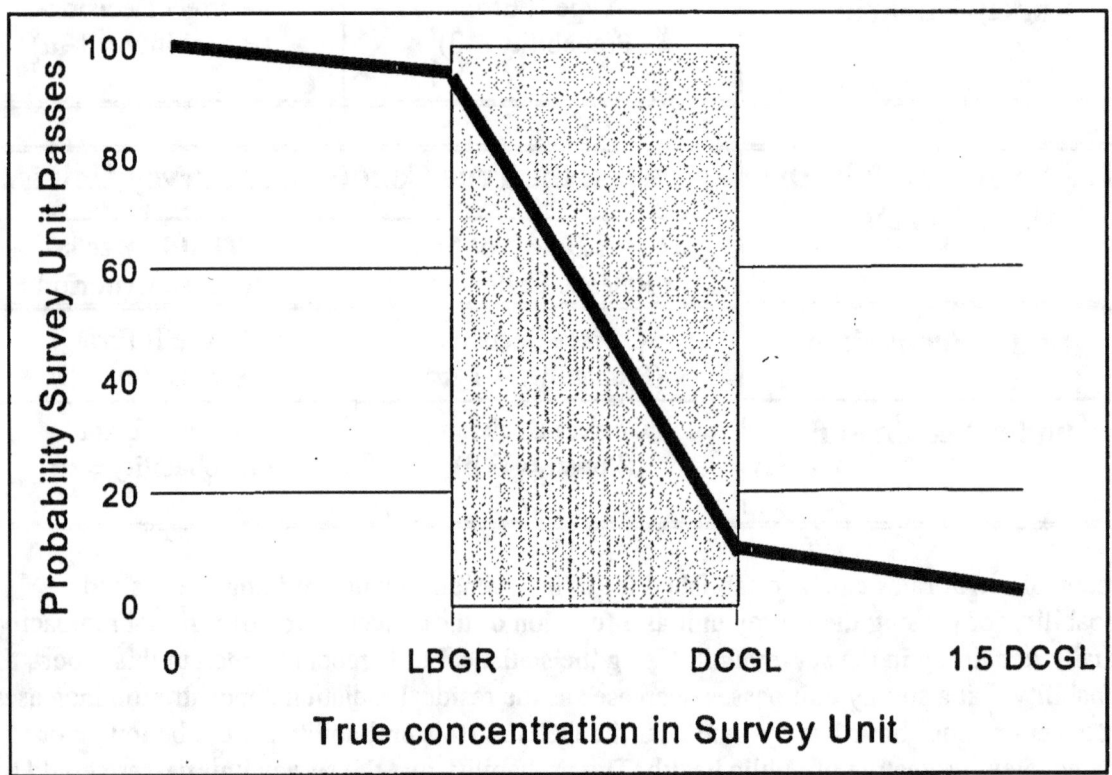

Figure 3.5 Example of Setting Acceptable Probabilities for Survey Unit Release

There is a relationship between α and β that is used in developing a survey design. When a fixed number of concentration measurements are made, increasing α generally decreases β, and decreasing α generally decreases β. Increasing the number of measurements allows either or both error rates to be decreased. Once the LBGR and DCGL$_W$ are specified, the number of measurements needed to meet the desired values of α and β from the statistical test can be estimated using the estimated variance of the measured concentration distribution. This is discussed further in Section 3.8. The technical details are given in Chapter 9.

Constructing a curve such as that in Figure 3.5 is equivalent to specifying the desired power curve for the statistical test that is used to analyze the final status survey data. The desired power curve for the statistical test is selected during the DQO process by specifying the desired values for α and β at the lower (LBGR) and upper (DCGL$_w$) boundary of the gray region. By definition, the power of a test at any specific concentration is the probability that the null hypothesis is rejected when that is the true concentration in the survey unit. In conducting a statistical test, the value of the test statistic is calculated, and compared to a critical value. The critical value depends only on α and the number of measurements, n. The actual power may larger or smaller than the desired power at any specific value of the assumed true concentration. Thus, Figure 3.5 *is* the desired power for Scenario A, but it is one minus the power for Scenario B. More information about the power of the tests, how it is calculated, and a procedure for comparing the desired power to the actual power is given in Chapter 10.

The critical value of the test statistic depends, explicitly, only on α and the number of measurements, n. One consequence of this, is that the Type I error rate has traditionally been given precedence in experimental designs. Often, α is set at an arbitrarily low value, without regard to the impact on increasing β. EPA (QA/G-9, 1995) recommends that a more balance approach be used, that the errors rates, α and β, be considered simultaneously, and that several different sets of values be examined before finalizing the survey design. This is part of the optimization process discussed in Section 3.8.

Another consequence of the explicit dependence of the critical value on α, is the practice of calculating p-values, or levels of significance. Recall that α is the probability that the null hypothesis is rejected when it is actually true. No data are needed to calculate this probability. However, once the data are obtained, one may calculate the probability that a data set as extreme as that observed would occur when the null hypothesis is true. Unless this calculated "p-value" is greater than α, the null hypothesis is rejected. Difficulties can arise when a p-value very close to α is calculated. There is a tendency, in this case, to believe that one has "just missed" the desired result. There is also a temptation to believe that if more data are taken, the p-value will fall on the "correct" side of α. Unfortunately, unless the survey is designed specifically to be performed in two stages, the p-value calculated following the second stage of data collection will no longer be the correct one for the original null hypothesis. Some specific ways to construct two-stage tests are referenced in Chapter 14.

The value of α is fixed during the DQO process so that the critical value of the test statistic will be an objective standard of comparison for the measured data. This is necessary so that a clear line between pass and fail is drawn, despite the measurement uncertainty. In setting the value of α, it can be useful to consider what level of discomfort will be felt about the decision if the p-value that is observed should fall a little bit to either side of it. The time to adjust α and β is during the DQO process, not after the data are taken.

As stated earlier, the acceptable probabilities for survey unit release, and the corresponding values of α and β, that are set as goals during in the DQO process should reflect the risk involved in making a decision error. The following are important considerations for this process:

- In radiation protection practice, the public health risk is modeled as a linear function of dose (BEIR, 1990). Therefore, a 10% change in dose results in a 10% change in risk. This situation is quite different from one in which there is a threshold. In the latter case, the risk associated with a decision error can be quite high, and a low decision error rate is desirable. When the risk is linear, higher error rates might be considered adequately protective at the boundaries of the gray region, especially since these errors are known to decrease as the concentration increases. One should consider, as part of the DQO process, the magnitude, significance, and potential consequences of decision errors at all concentration values. This is the purpose of the power curve.

- The DCGL itself has uncertainty associated with it. Since dose cannot be measured directly, dose pathway models are used. Many assumptions are made in converting doses to derived concentrations. To be adequately protective of public health, many models, especially screening models, are generally designed to guard against underestimating the dose that may be delivered by a given concentration of residual radioactivity. That is, the model assumptions tend to be such that the true dose delivered by residual radioactivity in the survey unit is very likely to be lower than that predicted by the model. Unfortunately, it is difficult to quantfy this. Nonetheless, it is probably safe to say that most models are conservative. This is an additional consideration that could support the use of higher acceptable error rates in some situations. The assumptions made in any model used to predict DCGLs for a site can be examined to determine if the use of site specific parameters result in large changes in the DCGLs, or whether a site-specific model should be developed rather than designing a survey around DCGLs that may be too conservative.

- The economic risk of requiring additional remediation when a survey unit already meets the release criterion may be highly non-linear. The costs will depend on whether the survey unit has already had remediation work performed on it, and the type of residual radioactivity present. There may be a threshold below which the remediation cost rises very rapidly. If so, a high probability of release is appropriate at that threshold value. This is primarily an issue for survey units that have a substantial likelihood of falling at or above the gray region for residual radioactivity. For survey units that are very lightly contaminated, or have been so thoroughly remediated that any residual radioactivity is expected to be far below the DCGL, smaller release probabilities may be tolerated, especially if final status survey sampling costs are a concern. Again, it is important to examine the probability that the survey unit passes over the entire range of possible residual radioactivity concentrations, below as well as above the gray region.

- Lower decision error rates may be possible if alternative sampling and analysis techniques can be used that result in higher precision. The same might be achieved with moderate increases in sample sizes. These alternatives should be explored before accepting higher design error rates. However, in some circumstances, such as high background variations, lack of a radionuclide specific technique, and/or radionuclides whose concentrations are very difficult and expensive to measure, error rates that are lower than the uncertainties in the dose estimates may be neither cost effective nor necessary for adequate radiation protection.'

3.7.2 Decision Errors for Elevated Areas

When the concern is to test the average residual radioactivity concentration, the actual surface area of the survey unit is immaterial except insofar as it should be consistent with that assumed in the dose pathway model used. It is only the distribution of the measured concentrations in the survey unit, its mean and its variance that are important. When the concern is finding isolated areas of elevated activity, the size of the survey unit must be explicitly taken into account. This is because the probability of discovering an elevated area depends on the sampling density, i.e., the distance between sampling locations.

From Section 3.5.4, the length (or spacing), L, of the systematic pattern is given by:

$$L = \sqrt{\frac{A}{0.866\ n}} \quad \textit{for a triangular grid}$$

and

$$L = \sqrt{\frac{A}{n}} \quad \textit{for a square grid}$$

where A is the area of the survey unit.

A computer code for determining the probability that an elliptically shaped elevated area would be missed by a systematic sampling grid was developed by Singer (1972). An elliptical area can be described by the length Λ, of its semi-major axis and its shape (the ratio of major and minor axis lengths), S. For a circle the length is simply the radius and the shape, S, is one. Figure 3.6 shows an example of a circular ($S = 1.0$) area with radius $L/2$ and an elliptical ($S = 0.2$) area with semi-major axis length L, compared to both square and triangular sampling grids with spacing L.

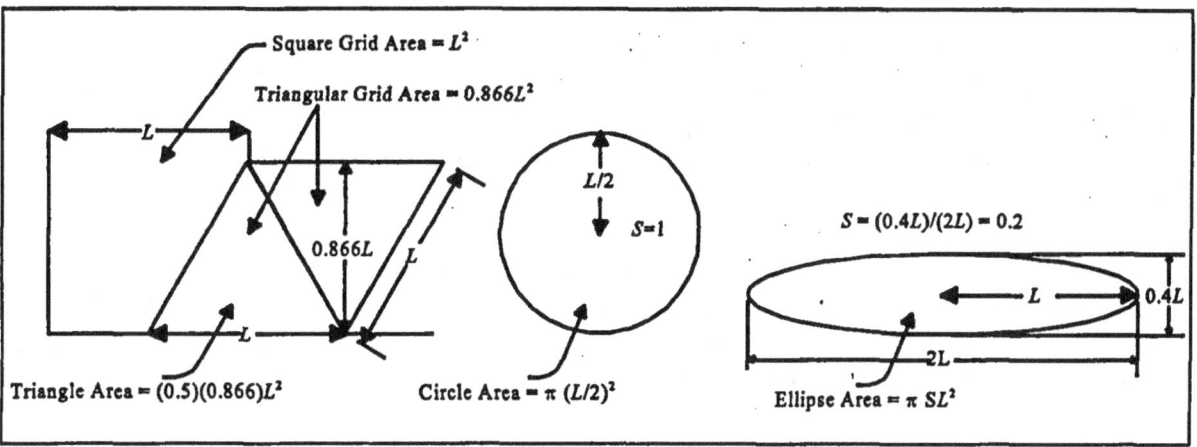

Figure 3.6 Circular and Elliptical Areas Relative to the Sample Grid

Singer's computer code, ELIPGRID, has been improved and modified for use on personal computers by Davidson (see ORNL/TM-12774). This code, ELIPGRID-PC, was used to generate the data for Figure 3.7.

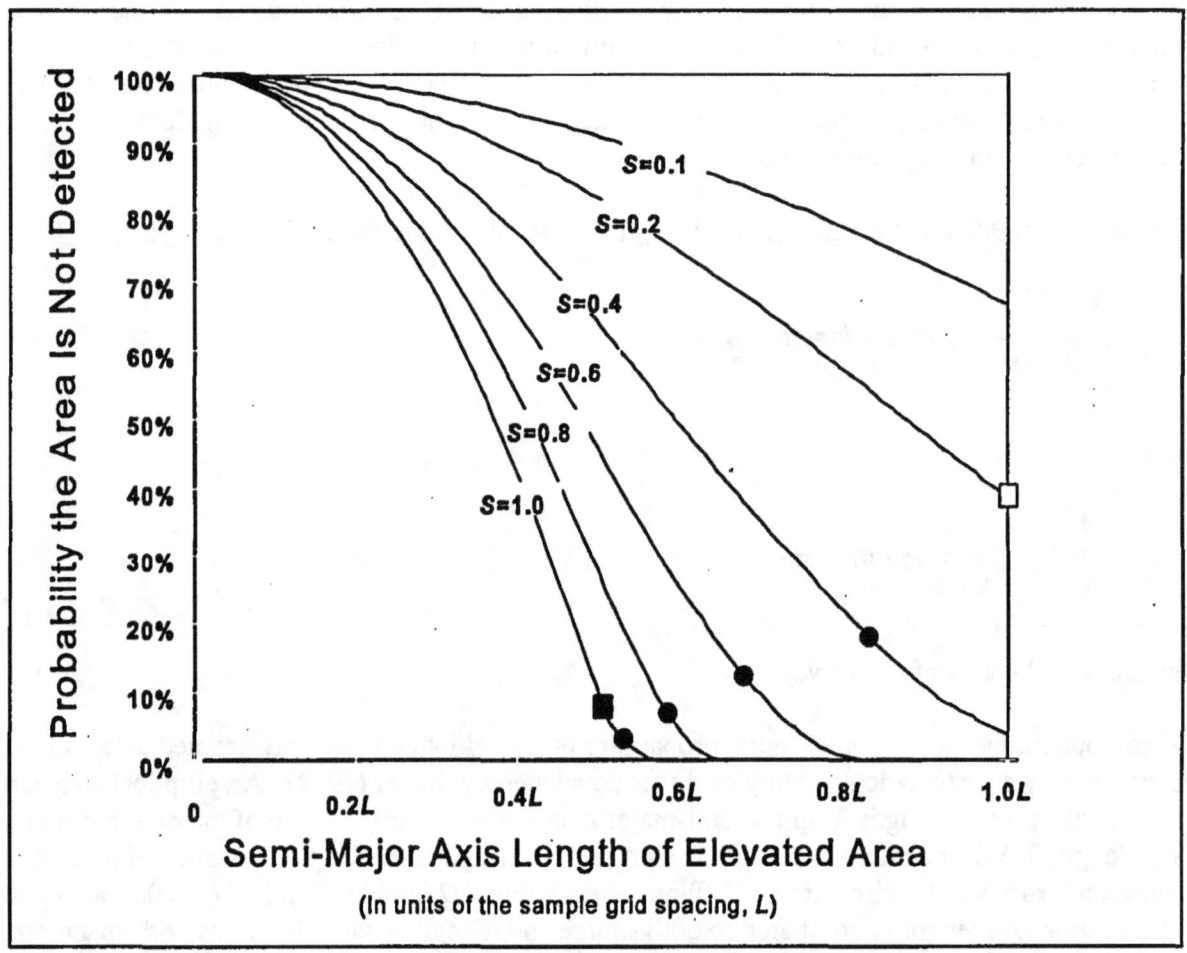

Figure 3.7 Probability an Elliptical Area Is Not Sampled on a Triangular Grid

This figure shows the probability that an elevated area of a given size and shape is not detected using a triangular sampling grid with spacing L. Note that the area of an ellipse of length Λ is $\pi S\Lambda^2$, so that for a given value of, an ellipse of length Λ with shape $S = 0.2$ has only one-fifth the area of a circle ($S = 1.0$) of the same length, i.e., radius. That is one reason, in addition to the area becoming longer and thinner, that the probabilities increase as S decreases.

The black and white squares in Figure 3.7 correspond the circle ($S = 1.0$, $\Lambda = 0.5L$) and the ellipse ($S = 0.2$, $\Lambda = L$) shown in Figure 3.6. The probability is less than 10% that the circle would go undetected. The probability is about 40 percent that the ellipse would go undetected, even though its area ($0.628\,L^2$) is only slightly smaller than the area of the circle ($0.785\,L^2$).

The circles in Figure 3.7 correspond to elevated areas equal to the triangular grid area, $0.866L^2$. The probability of missing them is rather low unless the shape parameter is also very low.

PLANNING

In Figure 3.8, the probabilities of missing a circular elevated area with triangular and square systematic grids are compared. The square grid is only slightly less efficient than the triangular grid. It can be concluded that, in most cases, an elevated area of the same size as, or larger than, that defined by the sampling grid is likely to be discovered during the final status survey.

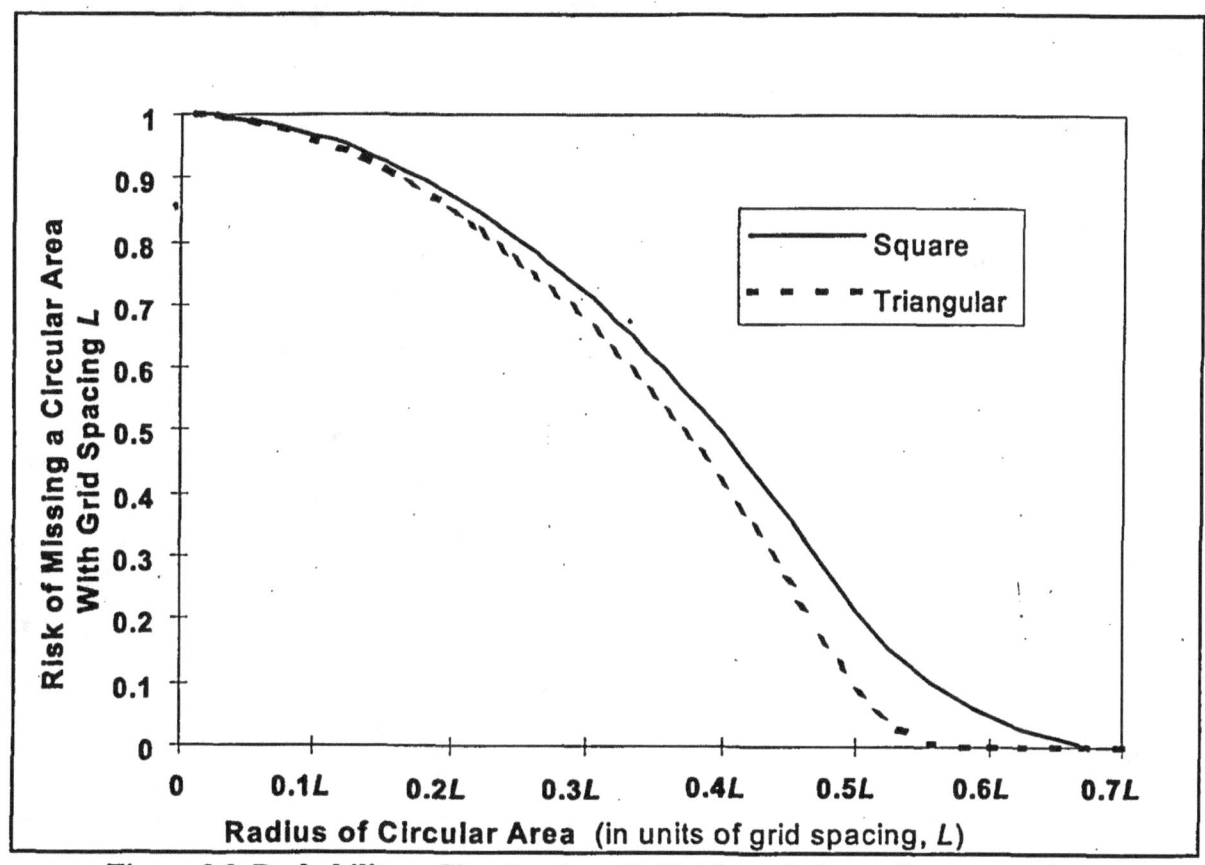

Figure 3.8 Probability a Circular Area Is Not Sampled on a Systematic Grid

3.8 Optimize the Design

The DQO process need be neither static nor sequential. Some of the activities involved may be taking place concurrently, and may be visited more than once. At any stage in the process, new information may be available that should then be incorporated into planning the final status surveys.

Optimization of the final status survey involves examining all of the factors that effect the decision errors and sample sizes so that costs and potential risks are balanced. The primary factors to be considered in optimizing the design for determining the mean concentration are the $DCGL_w$ and the measurement standard deviation. The estimate of the measurement standard deviation should include both the uncertainty in measurement process and any anticipated spatial and temporal concentration variations. The delineation and classification of survey units and reference areas can affect the spatial variability. Scan sensitivity is a primary consideration in optimizing the design to ensure no elevated areas remain in a survey unit. The Area Factor and

3-21   NUREG-1505

the scan MDC are the important parameters which can impact survey costs and uncertainty.

3.8.1 Optimizing the Design for the Mean Concentration

There are relationships between the measurement uncertainty, σ, the width of the gray region, Δ, the desired decision error limits (α and β) and the number of measurements needed to meet those limits. This is illustrated in Table 3.2 for the case when no reference area is needed (one-sample test). Table 3.3 is used when the survey unit is compared to a reference area (two-sample test), and lists the number of samples to be taken in each. The method used to generate these tables is discussed in Chapter 9.

Table 3.2 Number of Samples, N, Required in Survey Unit to Meet Error Rates α, and β, With Relative Shift Δ/σ, When Using the Sign Test

Δ/σ	α = 0.01				α = 0.05				α = 0.10				α = 0.25			
	β				β				β				β			
	0.01	0.05	0.10	0.25	0.01	0.05	0.10	0.25	0.01	0.05	0.10	0.25	0.01	0.05	0.10	0.25
0.1	4095	2984	2463	1704	2984	2048	1620	1018	2463	1620	1244	725	1704	1018	725	345
0.2	1035	754	623	431	754	518	410	258	623	410	315	184	431	258	184	88
0.3	468	341	282	195	341	234	185	117	282	185	143	83	195	117	83	40
0.4	270	197	162	113	197	136	107	68	162	107	82	48	113	68	48	23
0.5	178	130	107	75	130	89	71	45	107	71	54	33	75	45	33	16
0.6	129	94	77	54	94	65	52	33	77	52	40	23	54	33	23	11
0.7	99	72	59	41	72	50	40	26	59	40	30	18	41	26	18	9
0.8	80	58	48	34	58	40	32	21	48	32	24	15	34	21	15	8
0.9	66	48	40	28	48	34	27	17	40	27	21	12	28	17	12	6
1.0	57	41	34	24	41	29	23	15	34	23	18	11	24	15	11	5
1.1	50	36	30	21	36	26	21	14	30	21	16	10	21	14	10	5
1.2	45	33	27	20	33	23	18	12	27	18	15	9	20	12	9	5
1.3	41	30	26	17	30	21	17	11	26	17	14	8	17	11	8	4
1.4	38	28	23	16	28	20	16	10	23	16	12	8	16	10	8	4
1.5	35	27	22	15	27	18	15	10	22	15	11	8	15	10	8	4
1.6	34	24	21	15	24	17	14	9	21	14	11	6	15	9	6	4
1.7	33	24	20	14	24	17	14	9	20	14	10	6	14	9	6	4
1.8	32	23	20	14	23	16	12	9	20	12	10	6	14	9	6	4
1.9	30	22	18	14	22	16	12	9	18	12	10	6	14	9	6	4
2.0	29	22	18	12	22	15	12	8	18	12	10	6	12	8	6	3
2.5	28	21	17	12	21	15	11	8	17	11	9	5	12	8	5	3
3.0	27	20	17	12	20	14	11	8	17	11	9	5	12	8	5	3

Table 3.3 Number of Samples, *N/2*, Required in Both Reference Area and Survey Unit to Meet Error Rates α and β With Relative Shift Δ/σ, When Using the Wilcoxon Rank Sum Test

Δ/σ	α = 0.01 β				α = 0.05 β				α = 0.10 β				α = 0.25 β			
	0.01	0.05	0.10	0.25	0.01	0.05	0.10	0.25	0.01	0.05	0.10	0.25	0.01	0.05	0.10	0.25
0.1	5452	3972	3278	2268	3972	2726	2157	1355	3278	2157	1655	964	2268	1355	964	459
0.2	1370	998	824	570	998	685	542	341	824	542	416	243	570	341	243	116
0.3	614	448	370	256	448	307	243	153	370	243	187	109	256	153	109	52
0.4	350	255	211	146	255	175	139	87	211	139	106	62	146	87	62	30
0.5	227	166	137	95	166	114	90	57	137	90	69	41	95	57	41	20
0.6	161	117	97	67	117	81	64	40	97	64	49	29	67	40	29	14
0.7	121	88	73	51	88	61	48	30	73	48	37	22	51	30	22	11
0.8	95	69	57	40	69	48	38	24	57	38	29	17	40	24	17	8
0.9	77	56	47	32	56	39	31	20	47	31	24	14	32	20	14	7
1.0	64	47	39	27	47	32	26	16	39	26	20	12	27	16	12	6
1.1	55	40	33	23	40	28	22	14	33	22	17	10	23	14	10	5
1.2	48	35	29	20	35	24	19	12	29	19	15	9	20	12	9	4
1.3	43	31	26	18	31	22	17	11	26	17	13	8	18	11	8	4
1.4	38	28	23	16	28	19	15	10	23	15	12	7	16	10	7	4
1.5	35	25	21	15	25	18	14	9	21	14	11	7	15	9	7	3
1.6	32	23	19	14	23	16	13	8	19	13	10	6	14	8	6	3
1.7	30	22	18	13	22	15	12	8	18	12	9	6	13	8	6	3
1.8	28	20	17	12	20	14	11	7	17	11	9	5	12	7	5	3
1.9	26	19	16	11	19	13	11	7	16	11	8	5	11	7	5	3
2.0	25	18	15	11	18	13	10	7	15	10	8	5	11	7	5	3
2.25	22	16	14	10	16	11	9	6	14	9	7	4	10	6	4	2
2.5	21	15	13	9	15	11	9	6	13	9	7	4	9	6	4	2
2.75	20	15	12	9	15	10	8	5	12	8	6	4	9	5	4	2
3.0	19	14	12	8	14	10	8	5	12	8	6	4	8	5	4	2
3.5	18	13	11	8	13	9	8	5	11	8	6	4	8	5	4	2
4.0	18	13	11	8	13	9	7	5	11	7	6	4	8	5	4	2

The width of the gray region, Δ, is a parameter that is central to the nonparametric tests discussed in this report. It is also referred to as the *shift*. In this report, the gray region is always bounded from above by the DCGL$_w$ corresponding to the release criterion. The *Lower Boundary of the*

Gray Region (LBGR) is selected during the DQO process along with the target values for α and β, as discussed in Section 3.7.1. The width of the gray region, or shift, Δ, is equal to (DCGL − LBGR). The absolute size of the shift is actually of less importance than the *relative shift* Δ/σ, where σ is an estimate of the standard deviation of the measured values in the survey unit. The estimated standard deviation, σ, includes both the real spatial variability in the quantity being measured, and the precision of the chosen measurement method. The relative shift, Δ/σ, is an expression of the resolution of the measurements in units of measurement uncertainty. Expressed in this way, it is easy to see that relative shifts of less than one standard deviation, $\Delta/\sigma < 1$, will be difficult to detect. On the other hand, relative shifts of more than three standard deviations, $\Delta/\sigma > 3$, are generally easy to detect.

It is evident from Tables 3.2 and 3.3, that the number of measurements that will be required to achieve given error rates (α and β) depends entirely on the value of Δ/σ. Note also that the number of measurements required is symmetric in α and β. For example, if $\Delta/\sigma = 1$, $\alpha = 0.05$ and $\beta = 0.10$, then, from Table 3.1, the number of samples needed for the Sign test is 23. For the same value of Δ/σ, but with error rates reversed (i.e., $\alpha = 0.10$ and $\beta = 0.05$), the number of samples needed for the Sign test is again 23. Thus, these tables may be used to plan the number of measurements needed, regardless of whether Scenario A or Scenario B is used. It is only when the statistical test is *actually performed* on the measurement results that the distinction between α and β becomes important.

For fixed values of α and β, small values of Δ/σ result in large numbers of samples. It is desirable to design for $\Delta/\sigma > 1$ whenever possible. There are two obvious ways to increase Δ/σ. The first is to increase the width of the gray region by making LBGR small. The disadvantage is that the acceptable probability of the survey unit passing will be specified at this smaller LBGR. Thus, a survey unit will generally have to be lower in residual radioactivity to have a high probability of being judged to meet the release criterion. The second way to increase Δ/σ is to make σ smaller. One way to make σ small is by having survey units that are relatively homogeneous in the amount of measured radioactivity. This is an important consideration in selecting survey units that have both relatively uniform levels of residual radioactivity and also have relatively uniform background radiation levels. Measurements performed during scoping, characterization, and remedial action support surveys can be useful for determining an estimate of σ for the final status survey planning.

Another way to make σ small is by using more precise measurement methods. The more precise methods might be more expensive, but this may be compensated for by the decrease in the number of required measurements. One example would be in using a radionuclide specific method rather than gross radioactivity measurements for residual radioactivity that does not appear in background. This would eliminate the variability in background from σ, and would also eliminate the need for reference area measurements. On the other hand, the costs associated with performing additional measurements with an inexpensive measurement system may be less than the costs associated with fewer measurements of higher precision.

The effect of changing the width of the gray region and/or changing the measurement variability on the estimated number of measurements (and cost) can be investigated using Table 3.1 and 3.2. Generally, the design goal should be to achieve Δ/σ values between one and three. The number of samples needed rises dramatically when Δ/σ is smaller than one. Conversely, little is usually gained by making Δ/σ larger than about three. If Δ/σ is greater than three or four, one can take advantage of the measurement precision available by making the width of the gray region smaller. It is even more important, however, that overly optimistic estimates for σ be avoided. The consequence of taking fewer samples than are needed given the actual measurement variations will be increased error rates, leading to either unnecessary remediations (Scenario A) or improper survey unit release (Scenario B).

On the other hand, a smaller number of samples may still result in acceptable error rates. When Δ/σ is small, and the number of samples is large, a modest increase in the acceptable error rates may result significant reduction in the number of samples required. Given the other uncertainties involved, the cost savings may justify larger acceptable error rates. The advantage of the optimization step of the DQO process is that several alternatives can be explored on paper before time and resources are committed.

One consideration in setting the error rates are the health risks associated with releasing a survey unit that might actually contain residual radioactivity in excess of the DCGL. If a survey unit did exceed the DCGL, the first question that arises is "How much above the DCGL is the residual radioactivity likely to be?" Figures 3.9 through 3.12 can be used to estimate this.

These figures show the probability of the survey unit passing the statistical tests as a function of the true concentration of residual radioactivity in the survey unit. Figures 3.9 and 3.10 are for the one-sample Sign test, under Scenario A and B, respectively. Figures 3.11 and 3.12 are for the two-sample WRS test, under Scenario A and B, respectively. In these figures, the black-colored curves are those for $\alpha = 0.01$, the white-colored curves are those for $\alpha = 0.10$, and the gray-colored curves are those for $\alpha = 0.25$. For each value of α, survey unit sample sizes of 10, 15, 20, 30, 50 and 100 are shown. Note that in Scenario A, α is the probability that the survey unit passes when the concentration is equal to the $DCGL_w$. In Scenario B, α is the probability that the survey unit passes when the concentration is equal to the LBGR.

For example, if the $DCGL_w$ is 1.0, the LBGR is 0.5, σ is 1.0, $\alpha = 0.05$ and $\beta = 0.05$, then $\Delta/\sigma = 0.5$ and Table 3.2 indicates that 89 samples would be required. If $\alpha = 0.1$ and $\beta = 0.1$, then only 54 samples are required. How likely is it that a survey unit with residual radioactivity 50% higher than the $DCGL_w$ would pass? A concentration 50% higher than the $DCGL_w$ is 1.5, which is the same as the $DCGL_w + 0.5\sigma$. For the Sign test in Scenario A, Figure 3.9 (second white curve from the left) shows that the probability of the survey unit passing is near zero for a concentration of 1.5 when $\alpha = 0.1$ and the sample size is 50. While a survey unit with residual radioactivity equal to the $DCGL_w$ might have a 10% chance of being released, a survey unit at the $DCGL_w + 0.5\sigma$ has almost no chance of being released. On the other hand, a survey unit with a residual radioactivity that is at 50% of the $DCGL_w$, i.e., 0.5, is at the $DCGL_w - 0.5\sigma$, and has a 90 % chance of being released. If the sample size were nearer 100, the leftmost white curve shows that this probability would increase to about 99%. Thus, if the cost of remediation below a concentration of 0.5 was very high, the larger sample size might be chosen, but with the objective of achieving $\alpha = 0.10$ and $\beta = 0.01$.

A similar result is obtained for Scenario B, where a concentration of $1.5 = LBGR + 1.0\sigma$. Figure 3.10 (second white curve from the right) shows that with $\alpha = 0.1$ and a sample size of 50, a survey unit with a concentration level of LBGR + 1.0σ has less than a 1% chance of being released. Setting $\alpha = 0.01$ does not appear to appreciably increase this probability (second black curve from the right).

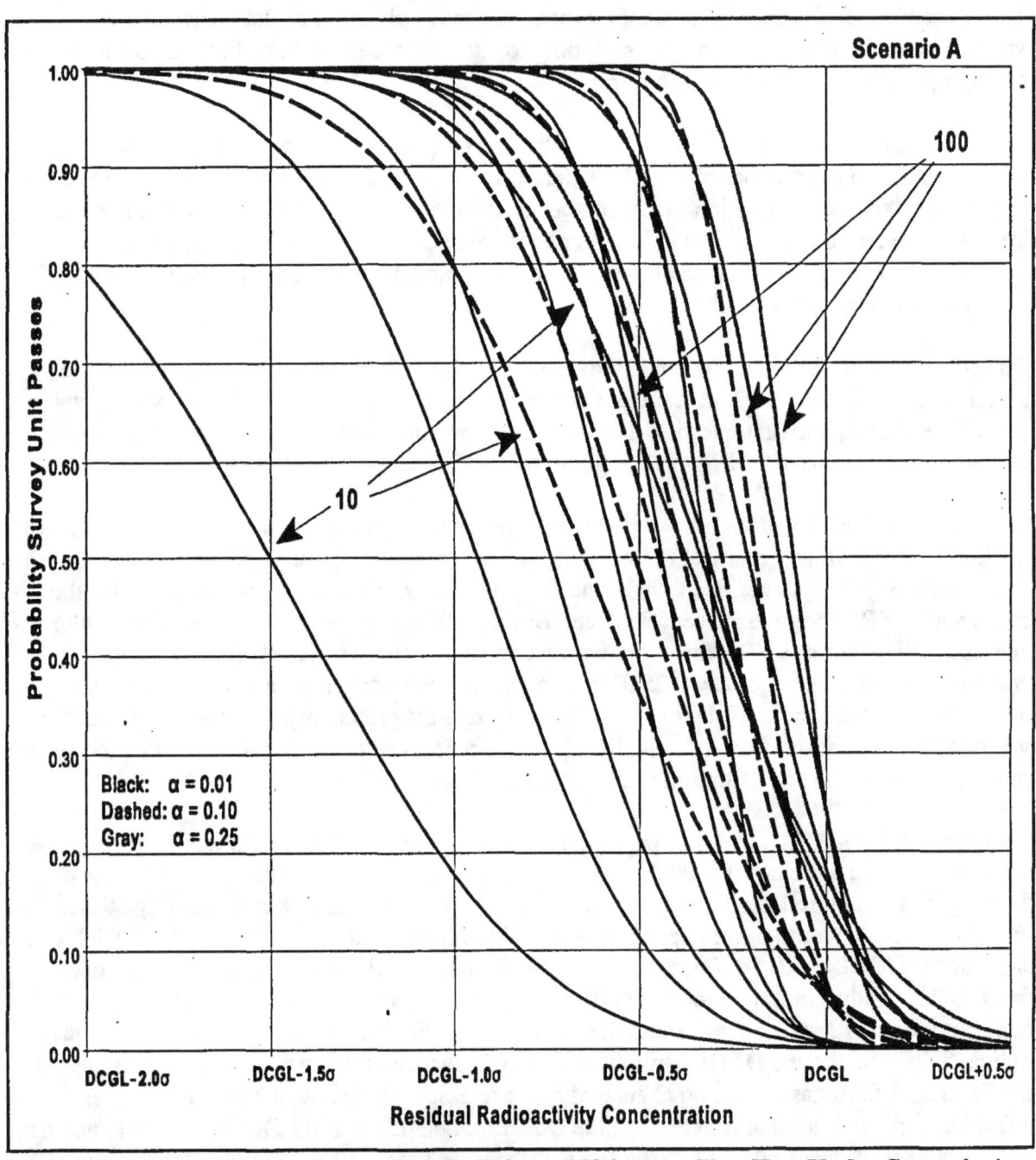

Figure 3.9 Probability a Survey Unit Is Released Using the Sign Test Under Scenario A

Using these figures, the probability that the survey unit passes over the entire range of possible residual radioactivity values, can be compared to the DQOs as expressed, for example, in Figure 3.5. In this way, the sample design can be optimized, taking into account the risks and costs associated with a decision error. The construction of the curves is discussed further in Chapter 10.

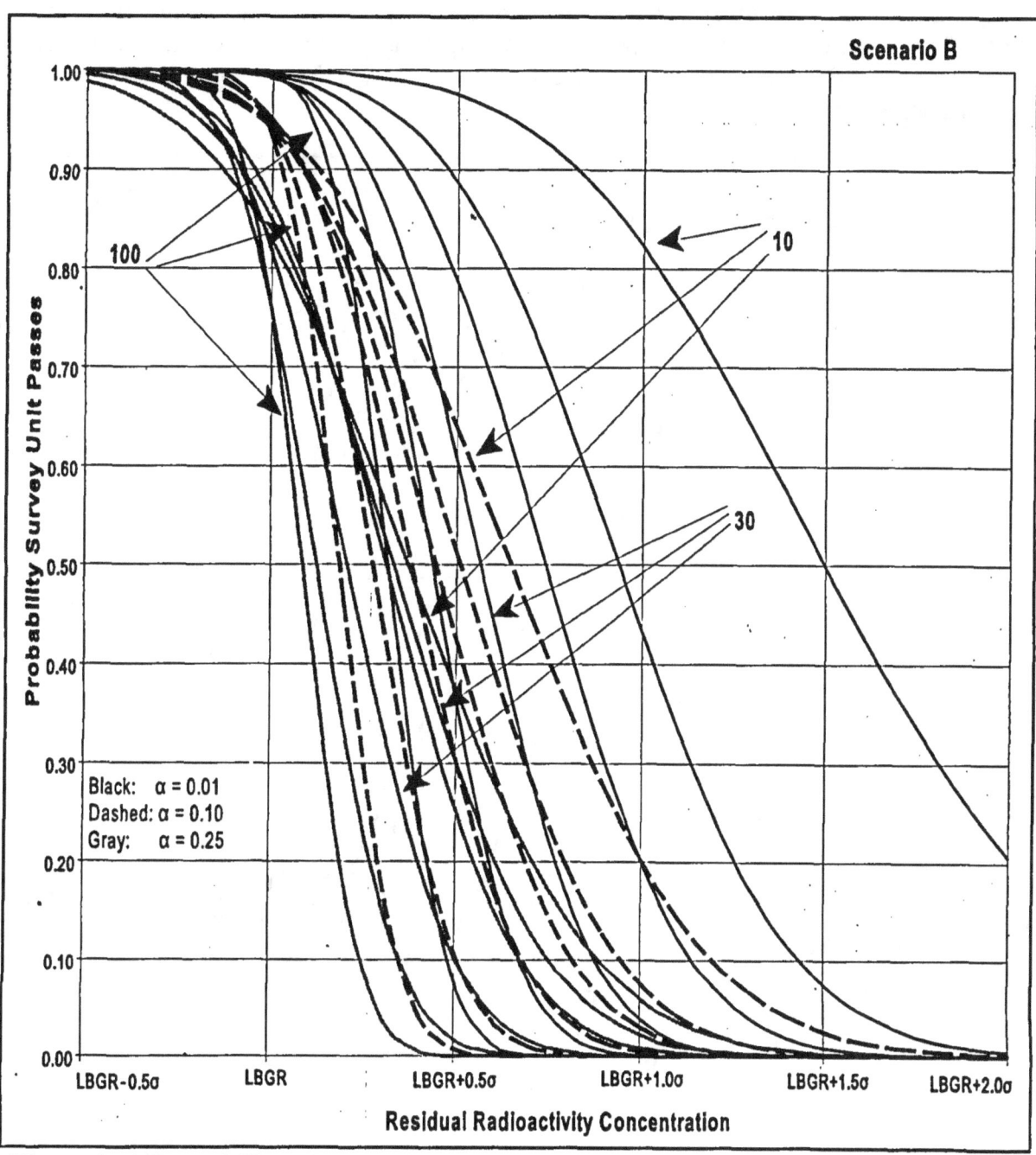

Figure 3.10 Probability a Survey Unit Is Released Using the Sign Test Under Scenario B

3.8.2 Optimizing the Design for Detecting Elevated Areas

As discussed in Section 2.2, one objective of the final status survey is to provide reasonable assurance that there are no small areas of elevated residual radioactivity left within the survey unit that might cause the release criterion to be exceeded. However, it is inefficient to treat all survey units equally in this regard. During the process of survey unit classification, Class 1 survey units are identified as those with the potential for such elevated areas. Measurements and

NUREG-1505

sampling on a systematic grid, in conjunction with scanning, are used to assure that any small areas of elevated radioactivity that might remain within a Class 1 survey units will not produce a dose in excess of the release criterion. To accomplish this, an additional step in the survey design optimization is required.

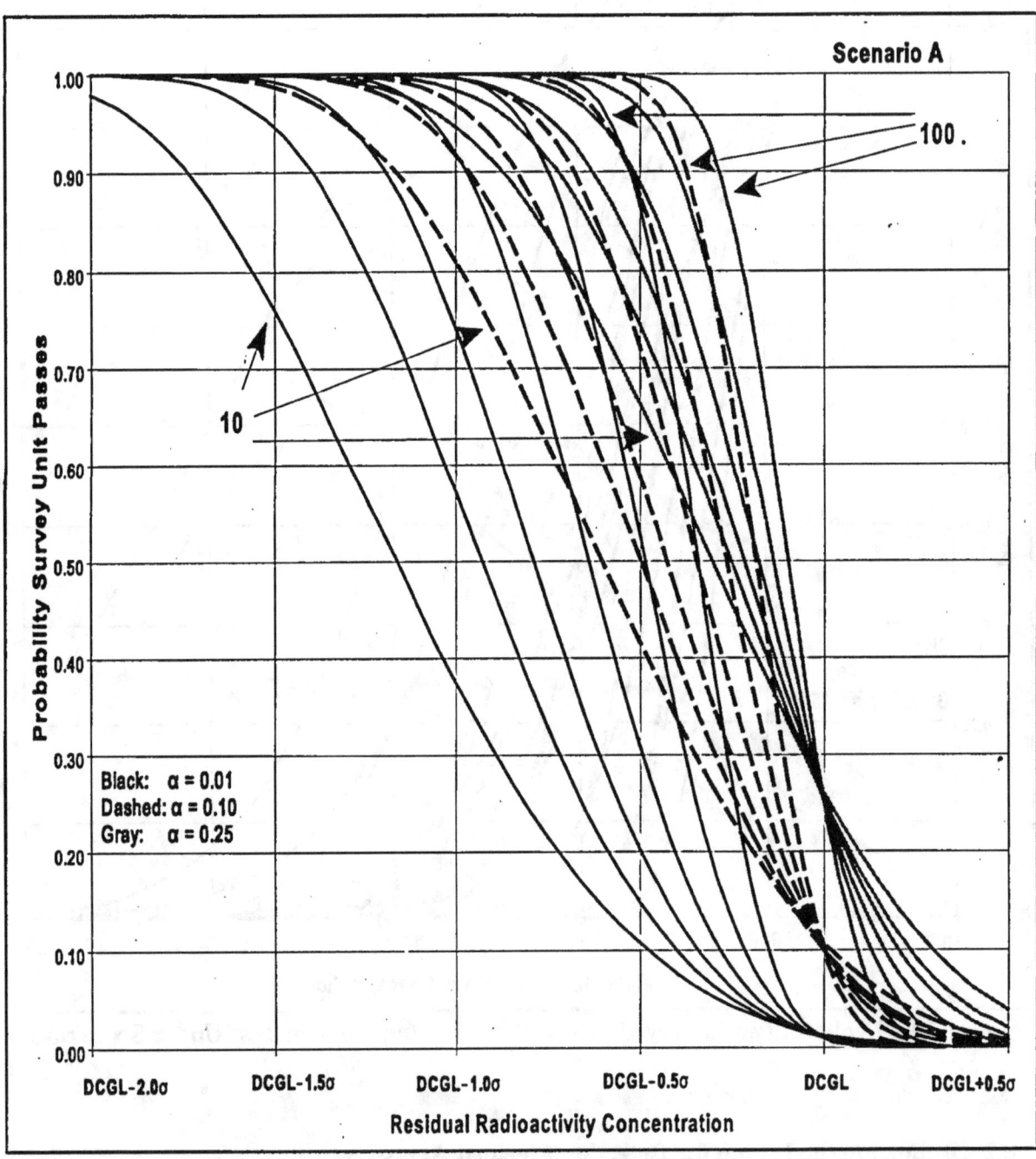

Figure 3.11 Probability a Survey Unit Is Released Using WRS Test Under Scenario A

The number of samples, N, to be taken on a random start systematic grid in a survey unit of area A, determines the spacing, L, between the samples (see Section 3.4.5). Corresponding to this spacing is the grid area delimited by neighboring sampling locations. This grid area is $0.866 L^2$ for a triangular grid and L^2 for a square grid.

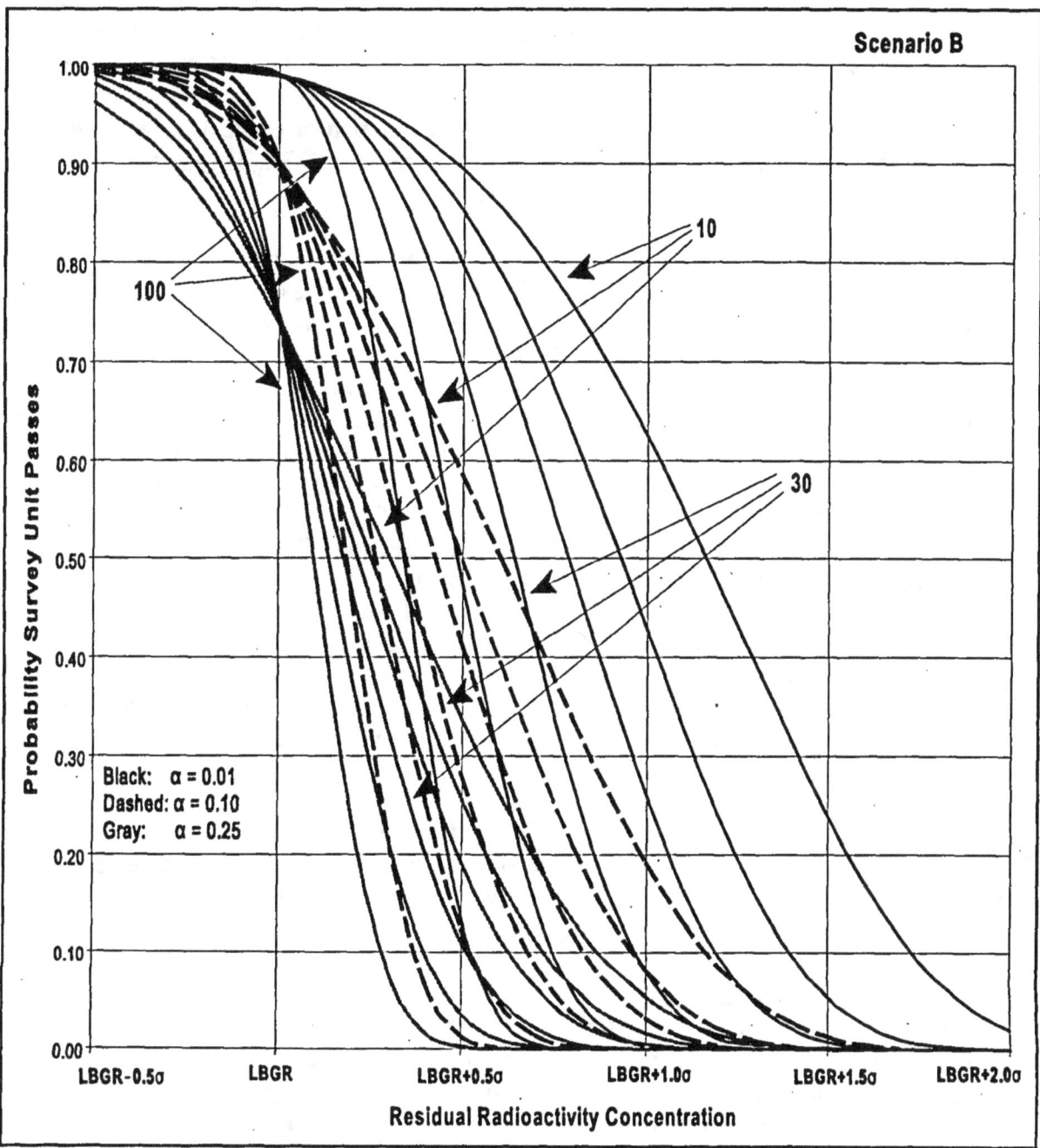

Figure 3.12 Probability a Survey Unit Is Released Using WRS Test Under Scenario B

A given concentration of residual radioactivity spread over a smaller area will, in general, result in a smaller dose. Thus, the $DCGL_{EMC}$ used for the elevated measurement comparison is usually larger than the $DCGL_W$ used for the statistical test. The amount of residual radioactivity that would have to exist within the grid area between sampling locations in order to exceed the guideline dose is a multiple, F_A, of the residual radioactivity derived concentration guideline level, $DCGL_W$. Values for the area factor, F_A, can be determined by comparing the dose conversion factor (DCF) obtained from the results of a pathway analysis under the scenario that a unit activity concentration of a given radionuclide is distributed uniformly across the survey unit to the DCF obtained when a unit concentration of that radionuclide is confined to the smaller grid area. For some radionuclides, especially those that deliver dose primarily via internal pathways, the dose is approximately proportional to inventory, and so the ratio of the $DCGL_{EMC}$ to the $DCGL_W$ is nearly proportional to the ratio of the survey unit area to the grid area. However, this may not be the case for radionuclides that deliver a significant portion of the dose via external exposure. The exact relationship between the $DCGL_{EMC}$ and the $DCGL_W$ is generally a complicated function of the dose modeling pathways.

The scanning procedure used for the survey unit should have a minimum detectable concentration (MDC) less than the $DCGL_{EMC}$. The $DCGL_{EMC}$ depends on the grid area which in turn depends on the spacing of the samples. Once a scanning technique is selected, the actual MDC can be compared to the $DCGL_{EMC}$. If the actual scan MDC is less than the $DCGL_{EMC}$, the survey design is complete. When the scanning method is sensitive enough to detect residual radioactivity concentrations at the $DCGL_{EMC}$, the combination of sampling and scanning will be sufficient to provide reasonable assurance that release criterion is met by any residual radioactivity remaining in the survey unit. Any area smaller than the grid area would require a residual radioactivity concentration within it larger than $DCGL_{EMC} = (F_A)(DCGL_W)$ in order to exceed the release criterion. Recall from the discussion in Section 3.7.2 that any area larger than the grid area is likely to be hit by a sampling location on the systematic grid at least once.

If the scanning MDC is greater than the $DCGL_{EMC}$, then the survey design must be modified. A larger $DCGL_{EMC}$ is generally obtained by decreasing the sampling grid area, i.e., adding more sampling locations to the grid. The number of additional sampling locations can be found by determining the area factor necessary to raise the $DCGL_{EMC}$ to a level detectable by scanning:

$$F_A = (Scan\ MDC)/(DCGL_W).$$

The sampling grid area, A_{MDC}, that corresponds to this area factor can be used to determine the sample size for the survey unit. Dividing the survey unit area, A_S, by the revised sampling grid area A_{MDC} yields the required sample size, $n_{MDC} = A_S/A_{MDC}$.

Thus, for Class 1 Survey Units, the number of samples may be driven more by the need to detect small areas of elevated activity than by the requirements of the statistical tests. This in turn will depend primarily on the sensitivity of available scanning instrumentation, the size of the area of elevated activity, and the dose model. For many radionuclides, scanning instrumentation is readily available that is sensitive enough to detect residual radioactivity concentrations at the $DCGL_{EMC}$ derived for the sampling grid of direct measurements used in the statistical tests. Where instrumentation of sufficient sensitivity is not available, the number of samples in the survey unit can be increased until the area between sampling points is small enough (and the resulting area factor is large enough) that $DCGL_{EMC}$ can be detected by scanning. For some radionuclides (e.g., ^3H), the scanning sensitivity is so low that this process would never terminate—i.e., the number of samples required could increase without limit. Thus, an important part of the DQO process is to determine the smallest size of an area of elevated activity that it is important to detect, A_{min}, and an acceptable level of risk, R_A, that it may go undetected. Figures 3.7 and 3.8 can be used for this purpose. The ELIPGRID-PC computer code (ORNL/TM-12774, 1994) can also be used to calculate these risks.

In this part of the DQO process, the concern is less with areas of elevated activity that are found than with providing adequate assurance that negative scanning results truly demonstrate the absence of such areas. In selecting acceptable values for A_{min} and R_A, maximum use of information from the historical site assessment and all surveys prior to the final status survey should be used to determine what sort of areas of elevated activity could possibly exist, their potential size and shape, and how likely they are to exist.

4 ANALYSIS OF FINAL STATUS SURVEY RESULTS: DATA QUALITY ASSESSMENT

Data Quality Assessment (DQA) is the scientific and statistical evaluation of data to determine if the data are of the right type, quality, and quantity to support their intended use (EPA QA/G-9, 1995). There are five steps in the DQA process:

(1) Review the Data Quality Objectives (DQOs) and sampling design.
(2) Conduct a preliminary data review.
(3) Select the statistical test.
(4) Verify the assumptions of the statistical test.
(5) Draw conclusions from the data.

4.1 Review the Data Quality Objectives (DQOs) and Sampling Design

During survey design, acceptable error rates are specified for the statistical tests, and the desired probability that a survey unit will pass the release criterion is charted against the amount of residual radioactivity that actually may be present in order to test the efficacy of a proposed design. During the interpretation of survey results, it is important to determine that the objectives of the design have been met. The first and most straightforward way to check this is to ascertain that the number of usable measurements meet the requirement of the statistical tests as outlined in Section 3.8.1. The sample standard deviation, s, should also be compared to the estimate of the measurement variability, σ, that was used to determine the number of samples required. The consequence of there being too few measurements, or of there being higher than expected data variability, is that the Type II error rate β will be larger than planned, and the power of the test to detect departures from the null hypothesis, $1 - \beta$, is reduced. In Scenario A this means that a survey unit that meets the release criterion has a higher probability of being incorrectly deemed *not* to meet it. In Scenario B this means that a survey unit that does *not* meet the release criterion has a higher probability of being incorrectly deemed to meet it. After examining the number of usable measurements and their variability, the retrospective power of the test can be determined using the methods of Chapter 10. This is not usually necessary when the null hypothesis is rejected since the Type I error rate, α, is fixed when the statistical test is performed using the actual number of usable measurements.

Since the occurrence of missing or unusable data can impact the Type II error rates, a reasonable allowance for such occurrences should be built into the planning process by adding more measurements to the sample sizes listed in Tables 3.2 and 3.3.

The power of the statistical tests will also be reduced if data variability is greater than that assumed during the survey planning. The number of measurements required to meet the acceptable error specified during the planning process will not be sufficient if σ was underestimated. As mentioned in Section 2.2.6, the overall data variability may be considered to consist of two more or less independent components, the component due to uncertainty in the measurement process, σ_{meas}, and the component due to spacial variability in the concentrations

4-1   NUREG-1505 being measured, $\sigma_{spatial}$. Spatial variability was discussed in Section 3.5.1. The overall variability is approximately $\sigma = \sqrt{\sigma_{meas}^2 + \sigma_{spatial}^2}$. If either standard deviation is one-third or less of the other, there is not much point in trying to reduce it further. If the smaller contributor were eliminated entirely, at most σ would be reduced by a factor of $\sqrt{9/10} \approx 0.95$, i.e., only about a 5% gain. Efforts should be directed at reducing the dominant component of the data variability.

The quality of data is critical to the successful execution of a survey. Even if the measurement uncertainty is dominated by the spatial variability, poorly calibrated instruments could lead to either improperly labeling an area as still contaminated or releasing it when, in fact, it is above the guidelines. For this reason, calibrations must be performed regularly with traceable standards; the inherent precision of the survey instrument must be evaluated to determine if it meets the needs of the survey plan. Energy responses of instruments must be known so that appropriate applications are made to different radiation fields. Replicate, reference, and blank measurements are also an integral part of the survey methodology. Comparisons of field measurement results to those of laboratory sample analyses forms an important quality control check.

Bounds on measurement uncertainties should be established in the planning process and regularly assessed throughout the measurement program. Uncertainties in the measurements add to the variance in distribution of data sets and should be taken into consideration when selecting parameters for the statistical tests and in the interpretation of results of these tests. Failure to adequately consider the effect of measurement errors could result in the added expense of additional measurements. In the worst case, inadequate control of the Type II statistical errors as determined from a retrospective power calculation, could invalidate the final survey results and require a re-survey. For this reason, it is better to plan the surveys cautiously:

- It is better to overestimate the potential data variability than to underestimate it.

- It is better to take too many samples than too few.

- It is better to overestimate minimum detectable concentrations (MDCs) than to underestimate them.

Further information on quality assurance for environmental data may be found in EPA QA/R-5 (1994), EPA QA/G-5 (1996), and ANSI/ASQC (1994)

4.2 Conduct a Preliminary Data Review

To learn about the structure of the data—identifying patterns, relationships, or potential anomalies—one can review quality assurance (QA) and quality control (QC) reports, prepare graphs of the data, and calculate basic statistical quantities.

Radiological survey data are usually obtained in units that have no intrinsic meaning relative to DCGLs, such as the number of counts per unit time. For comparison of survey data to DCGLs, the survey data from field and laboratory measurements should be converted to DCGL units.

4.2.1 Basic Statistical Quantities

Basic statistical quantities that should be calculated for the sample data set are the
- mean
- standard deviation
- median

The average of the data can be compared to the reference area average and the $DCGL_w$ to get a preliminary indication of the survey unit status. Where remediation is inadequate, this comparison may readily reveal that a survey unit contains excess residual radioactivity—even before applying statistical tests. For example, if the average of the data exceeds the $DCGL_w$ and the radionuclide of interest does not appear in background, then it is obvious that the survey unit does not meet the release criterion. On the other hand, if every measurement in the survey unit is below the $DCGL_w$, the survey unit will always pass the Sign test.

The value of the sample standard deviation is especially important. If too large compared to that assumed during the survey design, this may indicate an insufficient number of samples were collected to achieve the desired test power.

The median is the middle value of the data set when the number of data points is odd, and is the average of the two middle values when the number of data points is even. Thus 50% of the data points are above the median, and 50% are below the median. Large differences between the mean and the median would be an early indication of skewness in the data. This would also be evident in a histogram of the data.

Table 4.1 lists an example of concentration data taken in a reference area and survey unit. For this example, the quantity and units of measurement have been left arbitrary. Basic statistical quantities can be calculated simply by using one of the many widely available personal computer programs that perform data analysis. Table 4.2 shows the result of a "descriptive statistics" command applied to the data of Table 4.1 using a spreadsheet program. In addition to the mean, median and standard deviation, this table lists several other useful parameters such as the minimum, maximum, mode, range, skewness and kurtosis.

For the example survey unit, the mean is 1.15 and the median is 1.05. The sample standard deviation is 0.46. The difference between the median and the mean, divided by the sample standard deviation is sometimes used as a simple measure of skewness. In this case, $(1.15 - 1.05)/0.46 = 0.22$.

The coefficient of skewness is the average cubed difference from the mean divided by the standard deviation cubed. The sample estimate of skewness is $g_1 = m_3/m_2^{3/2}$, where

$$m_3 = \sum_{i=1}^{n}(x_i-\bar{x})^3/n \text{ , and } m_2 = \sum_{i=1}^{n}(x_i-\bar{x})^2/n. \quad m_2 = \sum_{i=1}^{n}(x_i-\bar{x})^2/n.$$

NUREG-1505

For a sample from a normal distribution, g_1 is approximately normal with mean zero and standard deviation $\sqrt{6/n}$. The sample skewness for the survey unit data is 0.96. This is nearly four times $\sqrt{6/90} = 0.26$, indicating that there is some positive skewness in this data set.

Table 4.1 Example Final Status Survey Data

Reference Area							Survey Unit					
Point No.	Data Value	Point No.	Data Value	Point No.	Data Value		Point No.	Data Value	Point No.	Data Value	Point No.	Data Value
1	1.1	31	1.9	61	0.8		91	1.2	121	0.7	151	0.8
2	1.3	32	1	62	1.1		92	1.4	122	1.9	152	1.1
3	0.7	33	0.7	63	0.6		93	0.8	123	1.3	153	1.2
4	0.7	34	1.9	64	0.8		94	0.6	124	1.4	154	0.7
5	1.6	35	1.0	65	1.2		95	1.4	125	0.5	155	1.4
6	1.0	36	0.6	66	0.8		96	2.9	126	1.0	156	1.6
7	1.1	37	1.5	67	1.0		97	0.9	127	1.3	157	0.4
8	0.7	38	1.1	68	0.9		98	0.8	128	0.6	158	0.6
9	0.9	39	0.9	69	1.5		99	0.8	129	1.3	159	1.6
10	1.8	40	0.9	70	0.8		100	1.6	130	1.5	160	0.7
11	0.9	41	0.8	71	1.2		101	1.6	131	1.4	161	1.0
12	0.7	42	1.1	72	1.1		102	1.2	132	1.3	162	1.0
13	1.1	43	0.9	73	0.6		103	1.2	133	0.8	163	1.8
14	1.1	44	1.2	74	1.0		104	2.5	134	1.5	164	1.3
15	0.9	45	1.2	75	0.9		105	1.9	135	0.8	165	1.5
16	1.5	46	1.0	76	1.0		106	1.9	136	1.1	166	0.8
17	1.0	47	1.3	77	0.6		107	0.9	137	1.1	167	1.5
18	0.8	48	0.9	78	0.9		108	0.9	138	1.0	168	0.9
19	0.6	49	0.8	79	1.0		109	0.8	139	1.1	169	0.9
20	1.1	50	1.7	80	0.8		110	1.0	140	1.6	170	0.8
21	0.7	51	0.7	81	0.6		111	1.7	141	1.5	171	1.5
22	0.6	52	1.0	82	1.2		112	1.5	142	0.8	172	1.0
23	0.9	53	0.8	83	1.2		113	2.1	143	0.7	173	0.7
24	0.6	54	1.0	84	1.3		114	2.0	144	0.6	174	1.1
25	1.5	55	0.5	85	1.0		115	1.7	145	0.9	175	1.4
26	0.9	56	1.1	86	0.9		116	0.7	146	0.8	176	1.0
27	1.5	57	1.1	87	0.7		117	1.0	147	0.5	177	1.2
28	0.8	58	0.9	88	0.9		118	1.0	148	0.6	178	0.5
29	1.1	59	0.9	89	1.4		119	1.5	149	0.8	179	0.5
30	1.2	60	0.6	90	1		120	1	150	0.8	180	1.7

Table 4.2 Basic Statistical Quantities Calculated for the Data in Table 4.1

Reference		Survey Unit	
Mean	1.00	Mean	1.15
Standard Error	0.03	Standard Error	0.05
Median	1.00	Median	1.05
Mode	0.90	Mode	0.80
Standard Deviation	0.30	Standard Deviation	0.46
Sample Variance	0.09	Sample Variance	0.22
Kurtosis	0.93	Kurtosis	1.44
Skewness	0.95	Skewness	0.96
Range	1.4	Range	2.5
Minimum	0.5	Minimum	0.4
Maximum	1.9	Maximum	2.9
Count	90	Count	90

The kurtosis is the average fourth power of the difference from the mean divided by the variance squared. It is a measure of how "flat" the distribution is relative to normally distributed data. The sample estimate of kurtosis is $b_2 = m_4/m_2^2$, where

$$m_4 = \sum_{i=1}^{n} (x_i - \bar{x})^4/n, \text{ and } m_2 = \sum_{i=1}^{n} (x_i - \bar{x})^2/n = (n-1)s^2/n.$$

For a sample from a normal distribution, b_2 has mean three. The *sample coefficient of kurtosis* is $g_2 = b_2 - 3$. For very large samples from a normal distribution, g_2 has mean zero and standard deviation $\sqrt{24/n}$ (Snedecor and Cochran, 1980). The sample coefficient of kurtosis for the survey unit data is 1.44, and .Thus, the kurtosis appears to be significantly greater than zero. It is an indicator of how well the sample variance, s^2, estimates the true variance, σ^2, of the measurement data. The variance of the sample estimate of the variance is

$$\text{Var}(s^2) = \frac{2\sigma^4}{n-1}\left[1 + \frac{n-1}{2n}g_2\right]$$

The variance of s is given approximately by

NUREG-1505

$$\text{Var}(s) = \text{Var}(\sqrt{s^2}) \approx \frac{1}{4s}\,\text{Var}(s^2) = \frac{2\sigma^4}{4s(n-1)}\left[1+\frac{n-1}{2n}g_2\right] \approx \frac{s^3}{2(n-1)}\left[1+\frac{n-1}{2n}g_2\right],$$

where the propagation of error formula for the variance of the square root has been used (Taylor, 1990). For the example survey unit data with $s = 0.36$, $g_2 = 1.44$, and $n = 90$, we have

$$\text{Var}(s) \approx \frac{(0.46)^3}{2(89)}\left[1+\frac{89}{180}1.44\right]=(0.0973/178)(1.712)=0.000936.$$

The standard deviation of $s = \sqrt{\text{Var}(s)} = \sqrt{0.000936} = 0.0306$. Thus, one can estimate that, *very roughly*, $s = 0.46 \pm 0.03$.

An approximate $1-\alpha$ confidence interval for σ^2 may be obtained from

$$\left[\frac{(n-1)s^2}{1+g_2/n}\right]\bigg/ \chi^2_{n-1}(1-\alpha/2) < \sigma^2 < \left[\frac{(n-1)s^2}{1+g_2/n}\right]\bigg/ \chi^2_{n-1}(\alpha/2)$$

where $\chi^2_{n-1}(a)$ is the 100ath percentile of the chi-squared distribution with $n-1$ degrees of freedom (Box, 1953). Percentiles of the chi-squared distribution are given in Table A.5. With $s^2 = 0.22$, $g_2 = 1.44$, $n = 90$, $\chi^2_{89}(0.975) = 117$, and $\chi^2_{89}(0.025) = 64.8$, we find that with 95%

confidence $\left[\frac{(89)(0.22)^2}{1+1.44/90}\right]\bigg/ 117 < \sigma^2 < \left[\frac{(89)(0.22)^2}{1+1.44/90}\right]\bigg/ 64.8$ or $0.1647 < \sigma^2 < 0.2974$,

which implies that $0.406 < \sigma < 0.545$. This is not too different from the cruder estimate $s = 0.46 \pm (2)(0.03) = 0.46 \pm 0.06$, using a 2σ interval about the mean.

Examining the minimum, maximum, and range of the data may provide additional useful information. The minimum of the example survey unit data is 0.4 and the maximum is 2.9, so the range is $2.9 - 0.4 = 2.5$. This is 5.4 times the standard deviation of 0.46. Figure 4.1 shows that is well within the expected spread of values for this ratio, which is sometimes called the *studentized* range. These intervals were calculated for normally distributed data.

Absolute upper and lower bounds for the studentized range have been found by Thomson (1955). These bounds are fairly wide, but are useful in checking for errors in calculation. The upper bound is $\sqrt{2(n-1)}$. The lower bound is $2\sqrt{(n-1)/n}$, when n is even, and $2\sqrt{n/(n+1)}$, when n is odd.

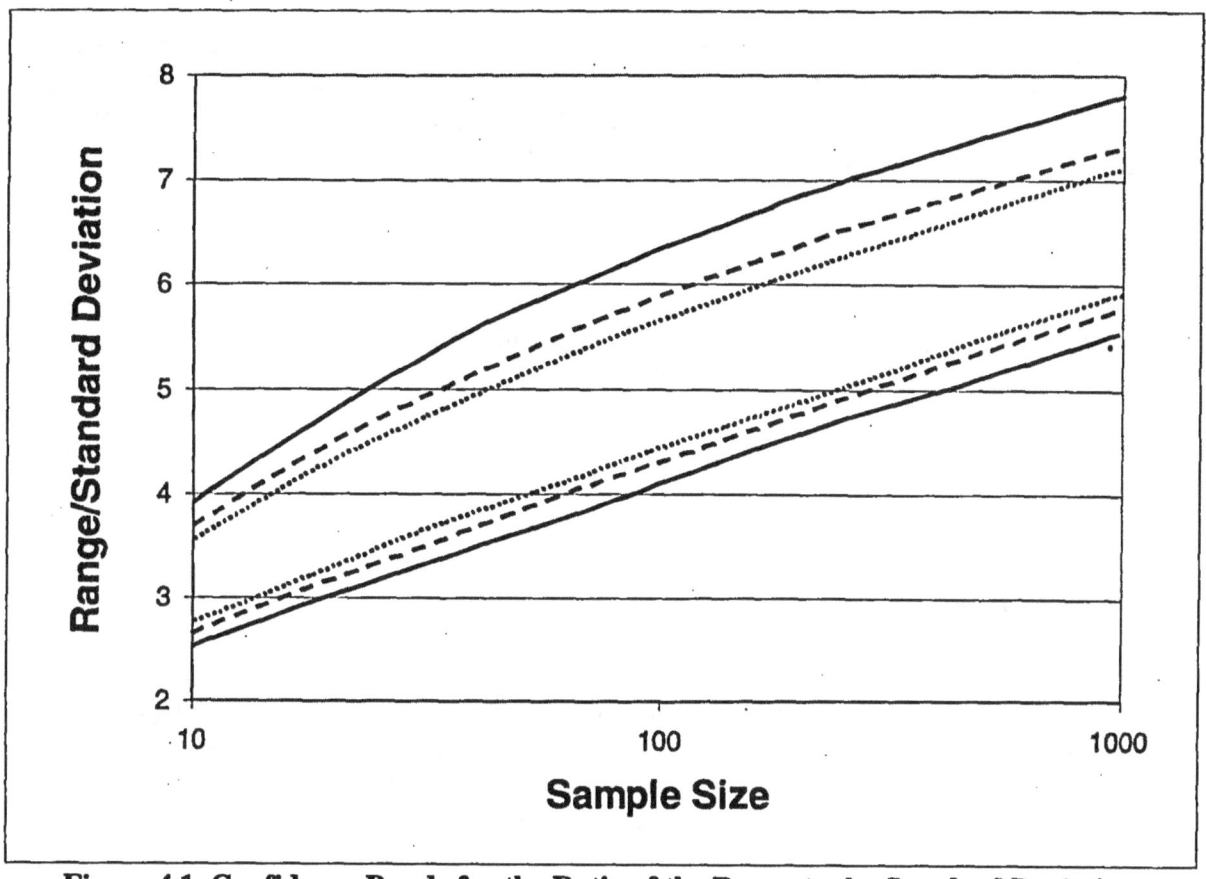

Figure 4.1 Confidence Bands for the Ratio of the Range to the Standard Deviation
Dotted: 90% Dashed: 95% Solid: 99%

Transformations are sometimes used bring data closer to a normal distribution, and decrease any dependence of the variance on the mean. A rule of thumb sometimes used is that if the ratio of the data maximum to the data minimum is less than 20, no data transformation is necessary to stabilize the variance of the data (EPA 600/4-90/013, 1990). For the example survey unit data, this ratio is 2.9/0.4 = 7.25.

Many of the "diagnostic checks" on the basic statistical quantities discussed in this section are based on comparing the values computed for the sample data distribution to those that would be expected if the data were normally distributed. When viewed as tests of normality, they are generally not very powerful, and are not suggested here for that purpose. As noted earlier, the nonparametric statistical tests used in this report do not assume the data are normally distributed. Rather, these checks are being used as exploratory techniques to alert the data analyst of any unusual features in the data.

4.2.2 Graphical Data Review

At a minimum, the graphical data review should consist of a posting plot and a histogram. Rank or Quantile plots are also useful diagnostic tools, particularly in the two-sample case, to compare the survey unit and reference area.

A *posting plot*, which is simply a map of the survey unit with the data values entered at the measurement locations, will reveal potential anomalies in the data, especially possible patches of elevated residual radioactivity. Even in a reference area, a posting plot can reveal spatial trends in background data that might affect the results of the two-sample statistical tests.

The survey unit data in Table 4.1 were taken on a square systematic grid in a rectangular survey unit. A simple posting plot is shown in the upper half of Figure 4.2. It is often useful to add some color coding of data values to aid in identifying patterns. In the lower half of Figure 4.2, darker shading was used for larger data values. The small slightly elevated region near 40 East and 20 North stands out more clearly when the shading is added.

If the posting plot reveals systematic spatial trends in the survey unit, the cause would need to be investigated. In some cases, such trends could be due to residual radioactivity, but may also be due to an inhomogeneous survey unit background. Other diagnostic tools for examining spatial data trends may be found in EPA Report QA/G-9 (1996). Geostatistical tools may also be useful in some cases (EPA 230/02-89-042, 1989a).

A *frequency plot* (or a histogram) is a useful tool for examining the general shape of a data distribution. This plot is a bar chart of the number of data points within a certain range of values. The frequency plot will reveal any obvious departures from symmetry, such as skewness or bimodality (two peaks), in the data distributions for the survey unit or reference area. Skewness or other asymmetry can impact the accuracy of the statistical tests. A data transformation (e.g., taking the logs of the data) can sometimes be used to make the distribution more symmetric. The statistical tests could then be performed on the transformed data. The interpretation of the results, however, can be more complex, since the quantity being tested is also transformed. For example, the mean of log-transformed data is the log of the geometric mean of the data, not the log of the arithmetic mean of the data.

The presence of two peaks in the survey unit frequency plot may indicate the existence of isolated areas of residual radioactivity. In some cases it may be possible to determine an appropriate background for the survey unit using this information. The interpretation of the data for this purpose will generally be highly dependent on site-specific considerations and should only be pursued after consultation with the responsible regulatory agency.

The presence of two peaks in the reference area frequency plot may indicate a mixture of background concentration distributions due to different soil types, construction materials, *etc.* The greater variability in the data due to the presence of such a mixture will reduce the power of the statistical tests to detect an adequately remediated survey unit. These situations should be avoided whenever possible by carefully matching the reference areas to the survey units, and choosing survey units with homogeneous backgrounds.

A major concern in constructing a histogram or frequency plot is the bin width, i.e. the range of concentration values over which the data are grouped and counted. If the bin width is too narrow, there will be too much spurious detail in the plot. If the bin width is too wide, too much detail is lost. A useful rule of thumb is to calculate the bin width by rounding down the quantity $3.5sn^{-1/3}$, where n is the number of data points, and s is the sample standard deviation (Scott, 1979). An example is shown in Figure 4.3 using the example survey unit data. In this example,

$3.5sn^{-1/3} = 3.5(0.46)(90)^{-1/3} = 3.5(0.32)(0.22) = 0.354$, which was rounded down to 0.3. The resulting histogram is shown as Figure 4.3a. For comparison, histograms constructed using bin widths of 0.2 (Figure 4.3b) and 0.1 (Figure 4.3c) are also shown.

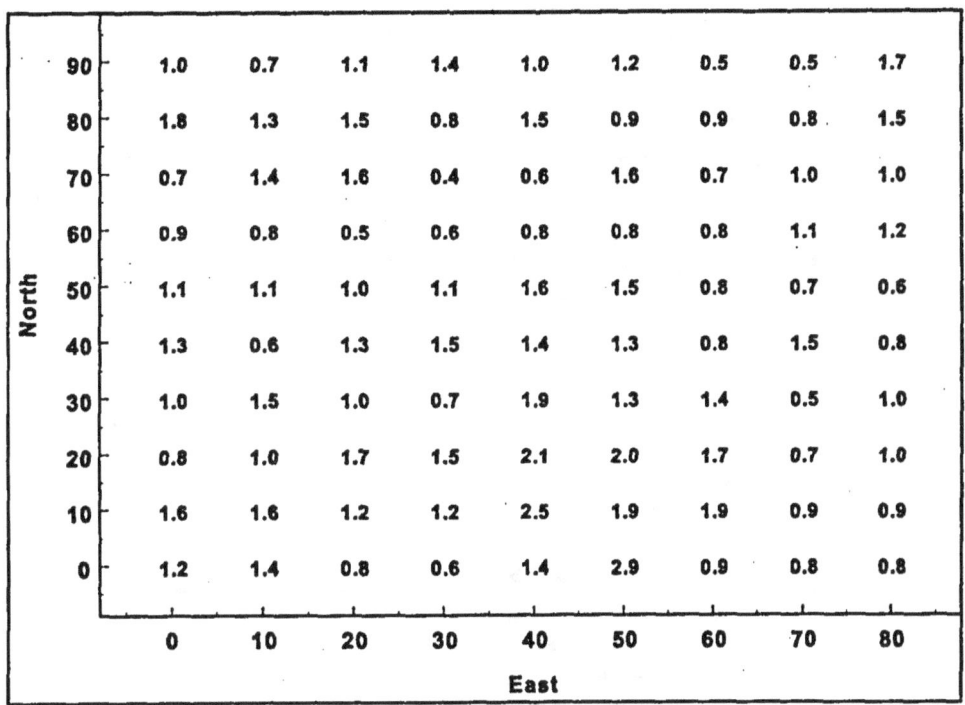

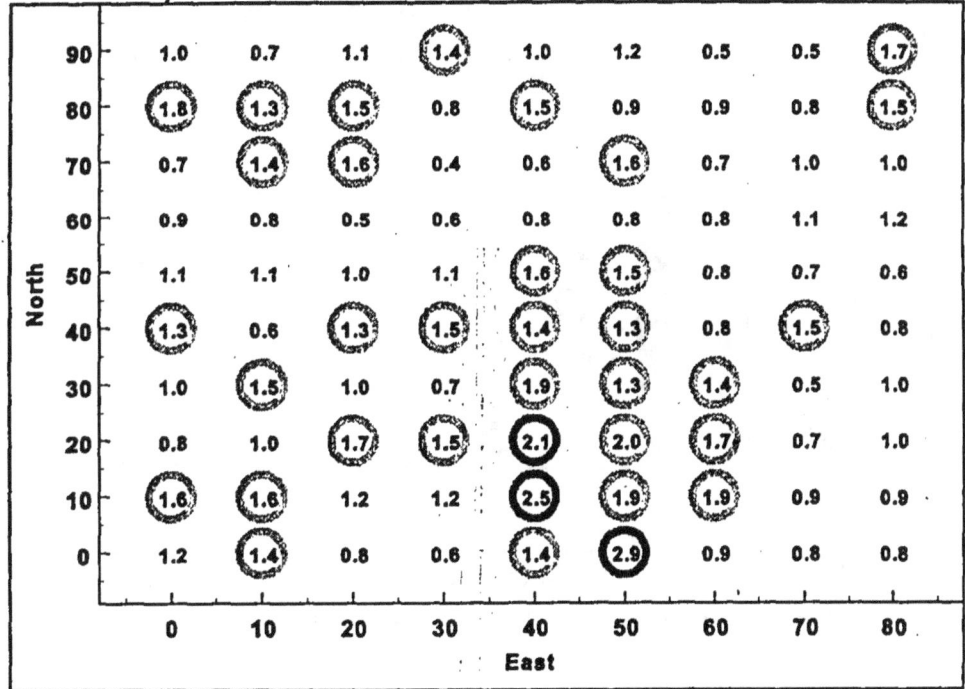

Figure 4.2 Example of a Posting Plot

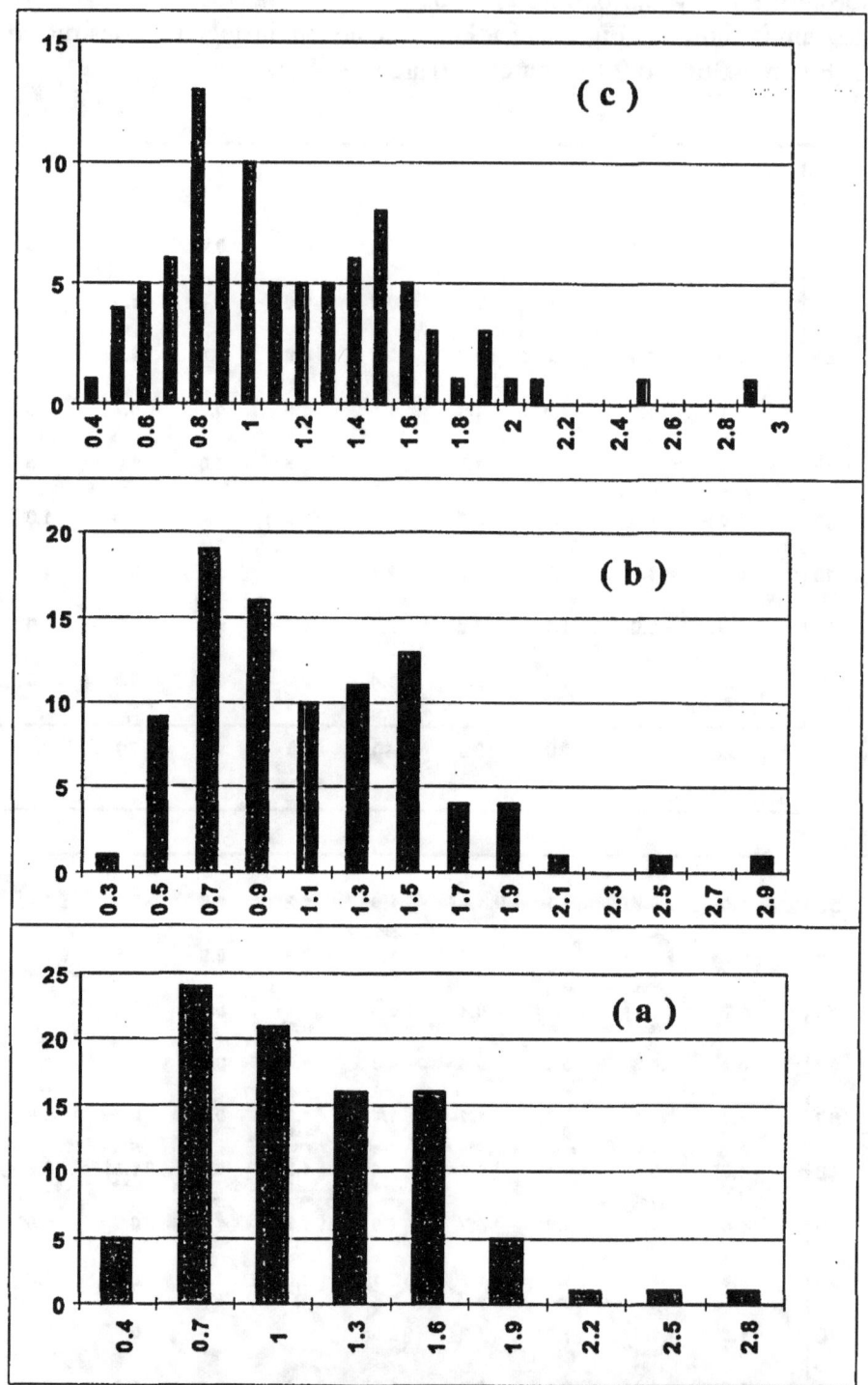

Figure 4.3 Frequency Plots of Example Survey Unit Data
with bin width 0.1 (c-top) and 0.2 (b-middle) and 0.3 (a-bottom)

A *ranked data plot* indicates the amount of data falling within a given range of values. The first step in constructing a ranked data plot is to sort the data in increasing order. The data are then assigned the number corresponding to their position in the list. The ranking of the example data from Table 4.1 is shown in Table 4.3.

Table 4.3 Ranks of the Example Data

Reference Area							Survey Unit					
Rank	Data	Rank	Data	Rank	Data		Rank	Data	Rank	Data	Rank	Data
1	0.5	31	0.9	61	1.1		1	0.4	31	0.9	61	1.4
2	0.6	32	0.9	62	1.1		2	0.5	32	0.9	62	1.4
3	0.6	33	0.9	63	1.1		3	0.5	33	0.9	63	1.4
4	0.6	34	0.9	64	1.1		4	0.5	34	0.9	64	1.4
5	0.6	35	0.9	65	1.1		5	0.5	35	0.9	65	1.4
6	0.6	36	0.9	66	1.1		6	0.6	36	1.0	66	1.4
7	0.6	37	0.9	67	1.1		7	0.6	37	1.0	67	1.5
8	0.6	38	0.9	68	1.1		8	0.6	38	1.0	68	1.5
9	0.6	39	0.9	69	1.1		9	0.6	39	1.0	69	1.5
10	0.6	40	0.9	70	1.2		10	0.6	40	1.0	70	1.5
11	0.7	41	0.9	71	1.2		11	0.7	41	1.0	71	1.5
12	0.7	42	0.9	72	1.2		12	0.7	42	1.0	72	1.5
13	0.7	43	0.9	73	1.2		13	0.7	43	1.0	73	1.5
14	0.7	44	0.9	74	1.2		14	0.7	44	1.0	74	1.5
15	0.7	45	1.0	75	1.2		15	0.7	45	1.0	75	1.6
16	0.7	46	1.0	76	1.2		16	0.7	46	1.1	76	1.6
17	0.7	47	1.0	77	1.3		17	0.8	47	1.1	77	1.6
18	0.7	48	1.0	78	1.3		18	0.8	48	1.1	78	1.6
19	0.8	49	1.0	79	1.3		19	0.8	49	1.1	79	1.6
20	0.8	50	.1.0	80	1.4		20	0.8	50	1.1	80	1.7
21	0.8	51	1.0	81	1.5		21	0.8	51	1.2	81	1.7
22	0.8	52	1.0	82	1.5		22	0.8	52	1.2	82	1.7
23	0.8	53	1.0	83	1.5		23	0.8	53	1.2	83	1.8
24	0.8	54	1.0	84	1.5		24	0.8	54	1.2	84	1.9
25	0.8	55	1.0	85	1.5		25	0.8	55	1.2	85	1.9
26	0.8	56	1.0	86	1.6		26	0.8	56	1.3	86	1.9
27	0.8	57	1.0	87	1.7		27	0.8	57	1.3	87	2.0
28	0.8	58	1.1	88	1.8		28	0.8	58	1.3	88	2.1
29	0.9	59	1.1	89	1.9		29	0.8	59	1.3	89	2.5
30	0.9	60	1.1	90	1.9		30	0.9	60	1.3	90	2.9

The ranked data plots for this data are shown in Figure 4.4 and Figure 4.5. A small amount of data in a range will result in a large slope. A large amount of data in a range of values will result in a flatter slope. A sharp rise near the bottom or the top is an indication of asymmetry. In Figure 4.4, the is an indication of some slight asymmetry in the reference area data. There is stronger evidence of asymmetry in the survey unit data in Figure 4.5.

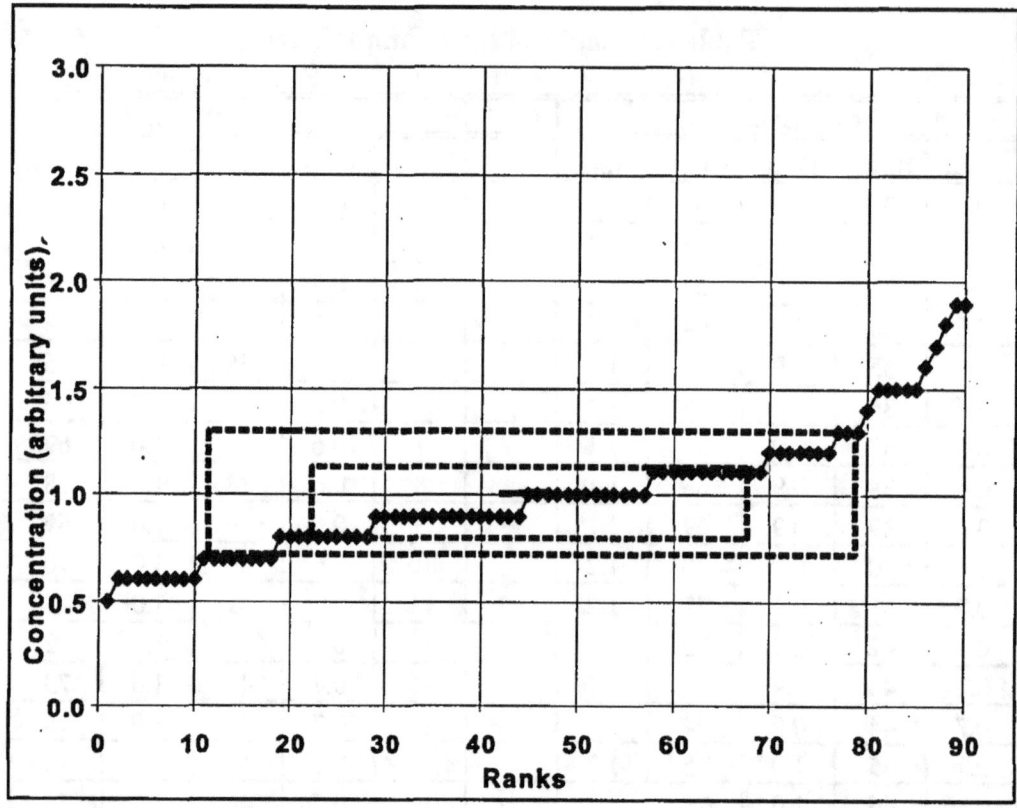

Figure 4.4 Ranked Data Plot for the Example Reference Area Data

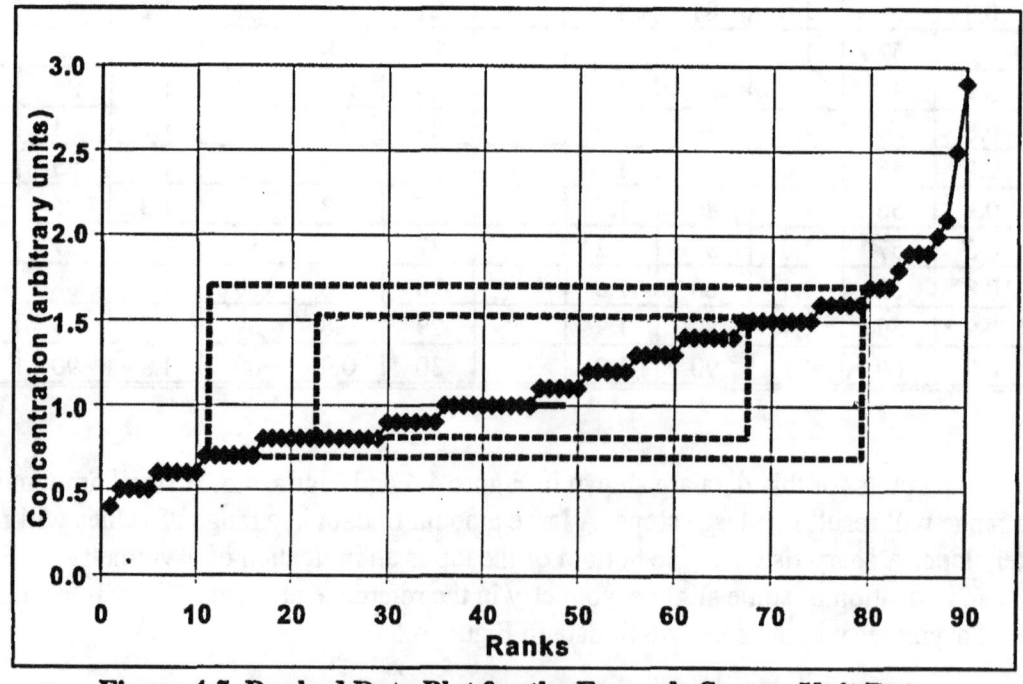

Figure 4.5 Ranked Data Plot for the Example Survey Unit Data

A Quantile plot is similar to a ranked data plot. It is constructed by ranking the data from smallest to largest, and simply plotting the data against the quantity: (rank−0.5)/(number of data points) rather than against the ranks. In this way, the percentage of data in various concentration ranges is easily found.

A useful aid to interpreting a ranked data or quantile plot is the addition of boxes containing the middle 50% and middle 75% of the data. These are shown as the dashed lines in Figure 4.4. The 50% box has its upper right corner at the 75th percentile and its lower left corner at the 25th percentile. These points are also called the quartiles. For the example survey unit data, these are 0.8 and 1.5, respectively. as indicated by the inner dashed box. They bracket the middle half of the data values. The 75% box has its upper right corner at the 87.5th percentile and its lower left corner at the 12.5th percentile. A sharp increase within the 50% box can indicate two or more modes in the data. Outside the 75% box, sharp increases can indicate outliers. The median (50th percentile) is indicated by the heavy solid line at the value 1.0, and can be used as an aid to judging the symmetry of the data distribution.

A Quantile-Quantile plot is valuable because it provides a direct visual comparison of the two data sets. If the two data distributions differ only in location (e.g., mean) or scale (e.g., standard deviation), the points will lie on a straight line. If the two data distributions being compared are identical, all of the plotted points will lie on the line $Y = X$. Any deviations from this would point to possible differences in these distributions. A Quantile-Quantile plot can be constructed to compare the distribution of the survey unit data with the distribution of the reference area data. If the number of data points is the same in both sets, the construction of the Quantile-Quantile plot is straightforward. This has already been done for the example data in Table 4.3. Simply plot each pair of measurements matched with the same rank, i.e. the survey unit measurement, Y, with rank R is plotted against the reference area measurement, X, with rank R. If the number of data points in the survey unit and reference area are not equal, the construction of the Quantile - Quantile plot will involve some numerical adjustments of the ranks. This and other useful techniques for exploratory data analysis are discussed in EPA QA/G-9 (1996).

The Quantile-Quantile plot for the example data is shown in Figure 4.6. The middle data point plots the median of the survey unit data against the median of the reference area data. That this point lies above the line $Y = X$, shows that the median of Y is larger than the median of X. Indeed, the most of the points lie above the line $Y = X$ in the region of the plot beyond a concentration value of about one. This is a sensitive indication that the distribution of the survey unit data is shifted toward values higher than the reference area distribution. As with the quantile plot, the addition of boxes containing the middle 50% and middle 75% of the data can be a useful aid to interpreting a quantile-quantile plot.

4.3 Select the Statistical Test

An overview of the statistical considerations important for final status surveys appears in Section 2.3, 3.7, and 3.8. The statistical tests recommended in this report for final status surveys are discussed in Section 2.4. The detailed instructions for applying these tests, with examples, appear in Chapters 5, 6, and 7.

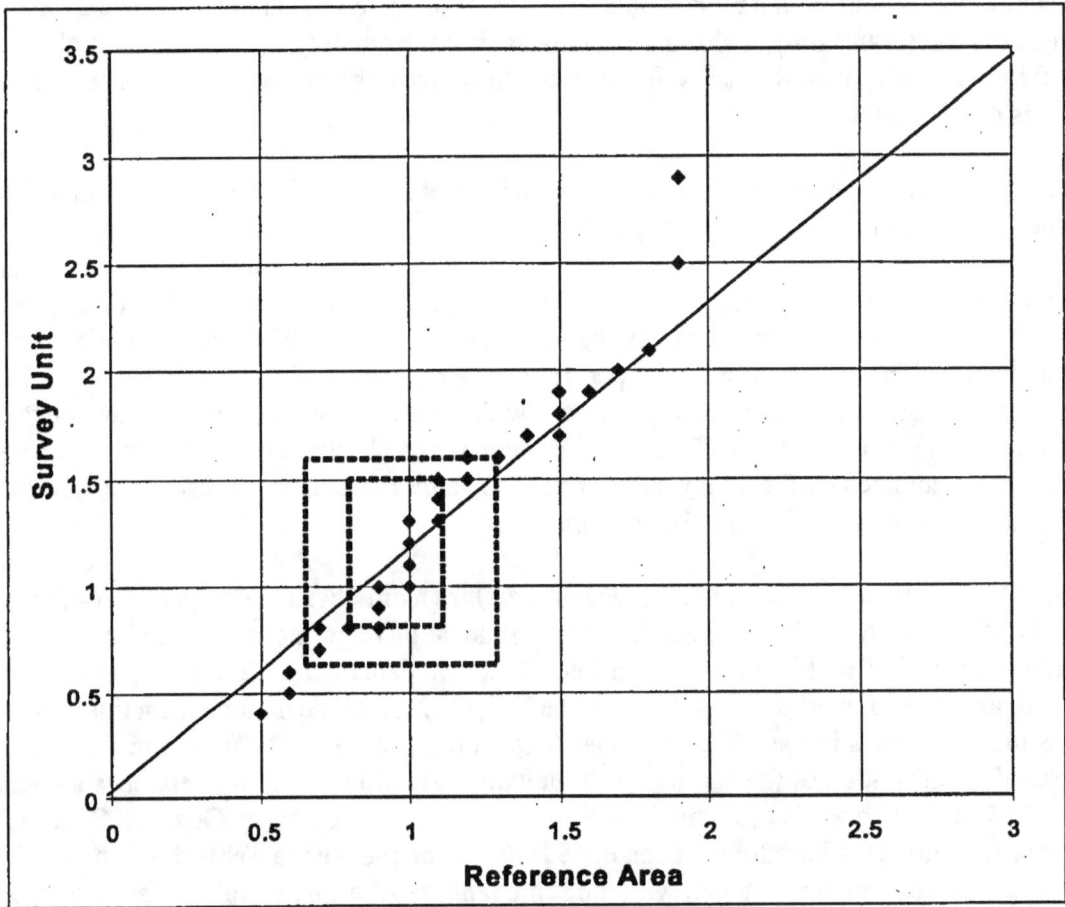

Figure 4.6 Example Quantile-Quantile Plot

The nonparametric statistical tests in this report are described as *one-sample* (Sign) and *two-sample* (WRS, Quantile) tests. Their application will depend upon the specific radionuclides under consideration, the concentration or surface activity limits for these radionuclides, and the comparison to background levels in the surrounding environment. Application of these techniques will also depend upon whether a gross dose or count rate survey is employed instead of spectrometric measurements for individual nuclides.

The one-sample tests are appropriate when there is no need to compare the survey unit with a reference area. The one-sample statistical test (Sign test) described in Chapter 5 can be used if the contaminant is not present in background and radionuclide-specific measurements are made. The one-sample test may also be used if the contaminant is present at such a small fraction of the $DCGL_w$ value as to be considered insignificant. In this case no provision for background concentrations of the radionuclide is made. Thus, the total concentration of the radionuclide is compared to the release criterion. This option should only be used if it is expected that ignoring the background concentration will not significantly affect the decision on whether or not the survey unit meets the release criterion. The advantage of ignoring a small background contribution is that no reference area is needed. This can simplify the final status survey considerably.

The two-sample WRS test (discussed in Chapter 6) should be used when the radionuclide of concern appears in background or if measurements are used that are not radionuclide specific.

The two-sample Quantile test discussed in Chapter 7 is used only when the null hypothesis of Scenario B is chosen.

The statistical tests recommended in this report are listed in Table 4.4. In every case, these tests are supplemented by the elevated measurements comparison (cf. Sections 2.6, 3.7.2, 3.8.2 and Chapter 8). Other statistical tests may be used provided that the data are consistent with the assumptions underlying their use, as discussed in the next section. The nonparametric tests generally involve fewer assumptions than their parametric equivalents. For example, the Student's t test may be used if the data distribution is consistent with the assumption of normality. If the data do not exhibit a normal distribution, the nonparametric tests will generally produce smaller decision error rates.

Table 4.4 Recommended Statistical Tests

Scenario	Reference Area	Test
A	Yes	Wilcoxon Rank Sum
A	No	Sign
B	Yes	Wilcoxon Rank Sum, Quantile
B	No	Sign

4.4 Verify the Assumptions of the Statistical Test

An evaluation should be made to determine that the data are consistent with the underlying assumptions of the statistical testing procedures used. Certain departures from these assumptions may be acceptable when given the actual data and other information about the study. Much of the information gained in the preliminary data review (Section 4.2) is directly applicable to verifying the assumptions of the statistical tests, and is a major reason for emphasizing their use.

A statistical test is called robust if it relatively insensitive to departures from its underlying assumptions. The nonparametric procedures described in this report were chosen because they are robust for the problem of testing the value of mean concentrations of residual radioactivity in a survey unit. In cases where the data distributions are extremely skewed, these tests may not detect limited areas with concentration much higher than the average in the survey unit. This is one reason for supplementing these tests with the elevated measurement comparison.

The nonparametric tests of Chapters 5, 6 and 7 assume that the data from the reference area or survey unit consist of independent samples from each distribution. The WRS test assumes that the reference area and survey unit data distributions are the same except for a possible shift in the mean.

Spatial dependencies that potentially affect the assumptions can be assessed using the posting plots. More sophisticated tools for determining the extent of spatial dependencies are also available (e.g., EPA QA/G-9, 1996). These methods tend to be complex and are best used with guidance from a professional statistician.

Asymmetry in the data can be diagnosed with a histogram or a ranked data plot. Data transformations can sometimes be used to minimize the effects of asymmetry.

One of the primary advantages of the nonparametric tests used in this report is that they involve fewer assumptions about the data than their parametric counterparts. If parametric tests are used, (e.g., Student's *t*-test), then any additional assumptions made in using them should be verified (e.g., testing for normality). These issues are discussed in detail in EPA QA/G-9 (1996).

Some alternative tests that may be considered in certain situations are discussed in Chapter 14. For example, if the data are symmetric, the one-sample WSR test is generally more powerful than the Sign test.

Table 4.5 Methods for Checking the Assumptions of Statistical Tests

Assumption	Diagnostic
Spatial Independence	Posting Plot
Symmetry	Histogram, Quantile Plot, Skewness
Data Variance	Sample Standard Deviation, Kurtosis
Power is Adequate	Retrospective Power Chart

4.5 Draw Conclusions From the Data

Perform the calculations required for the statistical tests and document the inferences drawn as a result of these calculations. The specific details for conducting the statistical tests are given in Chapters 5, 6 and 7.

In each survey unit, there are two types of measurements made: (1) direct measurements or samples at discrete locations and (2) scans. The statistical tests are only applied to the measurements made at discrete locations. When the data clearly show that a survey unit meets or exceeds the release criterion, the result is often obvious without performing the formal statistical analysis. Table 2.3 in Section 2.5 discussed those circumstances where a conclusion can be drawn from a simple examination of the data.

Sections 2.5.6 and 2.5.7 discuss the elevated measurement comparison (EMC) and the investigation levels that flag a locations for further study in order to determine whether the survey unit meets or exceeds the release criterion.

This report has been fairly explicit about the steps that should be taken to show that a survey unit meets release criteria. Less has been said about the procedures that should be used if at any point the survey unit fails. This is primarily because there are many different ways that a survey unit may fail the final status survey. The overall level of residual radioactivity may not pass the nonparametric statistical tests. Further investigation following the elevated measurement comparison may show that there is a large enough area with a concentration too high to meet the dose criterion. Investigation levels may have been flagged during scanning that indicate unexpected levels of residual radioactivity for the survey unit classification. It is impossible to enumerate all of the possible reasons for failure, their causes, and their remedies.

When a survey unit fails the release criterion, the first step is to review and confirm the data that led to the decision. Once this is done, the extent of the residual radioactivity is that caused the failure should be determined. Once the cause of failure has been remediated, determine the additional data, if any, needed to document that the survey unit meets the release criterion.

For example, a Class 2 survey unit passes the nonparametric statistical tests, but has several measurements on the sampling grid that exceed the $DCGL_w$. This is unexpected in a Class 2 area, and according to Table 2.4, these measurements are flagged for further investigation. Additional sampling confirms that there are several areas where the concentration exceeds the $DCGL_w$ This indicates that the survey unit was mis-classified. However, the scanning technique that was used was sufficient to detect residual radioactivity at the $DCGL_{EMC}$ calculated for the sample grid. No areas exceeding the $DCGL_{EMC}$ where found. Thus, the only difference between the final status survey actually done, and that which would be required for Class 1, is that the scanning may not have covered 100% of the survey unit area. In this case, it would be reasonable to simply increase the scan coverage to 100%. If no areas exceeding the $DCGL_{EMC}$ are found, the survey unit has, in effect, met the release criteria as a Class 1 survey unit.

A second example might be a Class 1 Survey unit which passes the nonparametric statistical tests, but which contains some areas that were flagged for investigation during scanning. Further investigation, sampling and analysis indicates one area is truly elevated. This area has a concentration that exceeds the $DCGL_w$ by a factor greater than the area factor calculated for its actual size. This area is remediated, and remediation control sampling shows that the residual radioactivity was removed, and no other areas were contaminated with removed material. It may be reasonable in that case, to simply document the original survey and the additional remediation data. It is not clear that further final status survey data would provide any useful information.

As a last example, consider a Class 1 area which fails the nonparametric statistical tests. Confirmatory data indicates that the average concentration in the survey unit does exceed the $DCGL_w$ substantially over a majority of its area. There would appear to be little alternative to remediation of the entire survey unit, followed by another final status survey.

These examples are meant to illustrative of the actions that may be necessary to secure the release of a survey unit that has initially failed to meet the release criterion. The DQO process should be revisited so that a plan can be made for attaining the original objective: to safely release the survey unit by showing that it meets the release criteria. Whatever data is necessary to meet this objective will be in addition to the final status survey data already in hand. All of the data, and only the data, necessary to meet the objective should be required.

5 SIGN TEST: CONTAMINANT NOT PRESENT IN BACKGROUND

The statistical test discussed in this section is used to compare each survey unit directly with the applicable release criterion. With only the set of survey unit measurements being analyzed, the Sign test used here is called a one-sample test. This section applies if

(1) radionuclide-specific measurements are made to determine the concentrations, and

(2) the background concentration of the radionuclide is negligible.

Otherwise, the methods of Chapter 6 and 7 should be used. Together the above conditions eliminate the need for a reference. The residual radioactivity concentrations in the survey unit are compared directly to the $DCGL_W$ value. The background concentration of the radionuclide need not be zero, but this background amount will be included with the residual radioactivity when analyzing the survey results. The amount that is considered negligible depends on the fraction of the $DCGL_W$ that it represents, and how much residual radioactivity is actually in the survey unit. The risk of the survey unit failing because the background concentration of the radionuclide is included with the residual radioactivity total should be weighed against the savings obtained by not having to make reference area measurements. Sites need not be contiguous areas, however the statistical test are generally applied to individual survey units that cover contiguous areas.

The Sign test is designed to detect uniform failure of remedial action throughout the survey unit. This test does not assume that the data follow any particular distribution, such as normal or log-normal. In addition to the Sign Test, the $DCGL_{EMC}$ for the Elevated Measurement Comparison (EMC)—described in Chapter 8—is compared to each measurement to ensure none exceeds the $DCGL_{EMC}$. If a measurement exceeds this $DCGL_{EMC}$, then additional investigation is recommended—at least locally—to determine the actual areal extent of the elevated concentration.

5.1 Introduction

The use of the Sign test in Scenario A and Scenario B is described in the next two sections, illustrated with example data. The same data will be used in both cases. We consider a survey unit that has been remediated, but may have some residual radioactivity. The $DCGL_W$ for the radionuclide in question has been determined to be 15.9. (The particular radionuclide and units of measurement are irrelevant to the example, and will be left arbitrary.) From data collected during the remediation, it is estimated that the standard deviation of the measurements in the survey unit made during the final status survey should be about 3.3.

During the DQO process it was agreed that the decision error rates should be both set equal to 0.05 initially, and determine if a reasonable survey design could meet these. It was estimated that the costs of additional remediation would be moderate down to concentrations of about 11.5, but would rise sharply below that. On the basis of these considerations, a chart of the desired probability that the survey unit passes was developed and is shown in Figure 5.1.

The lower bound of the gray region is 11.5, and the DCGL$_w$ is 15.9, so $\Delta = 15.9 - 11.5 = 4.4$. Since σ is estimated at 3.3, $\Delta/\sigma = 4.4/3.3 = 1.3$. From Table 3.2, for $\alpha = \beta = 0.05$, 21 samples are required.

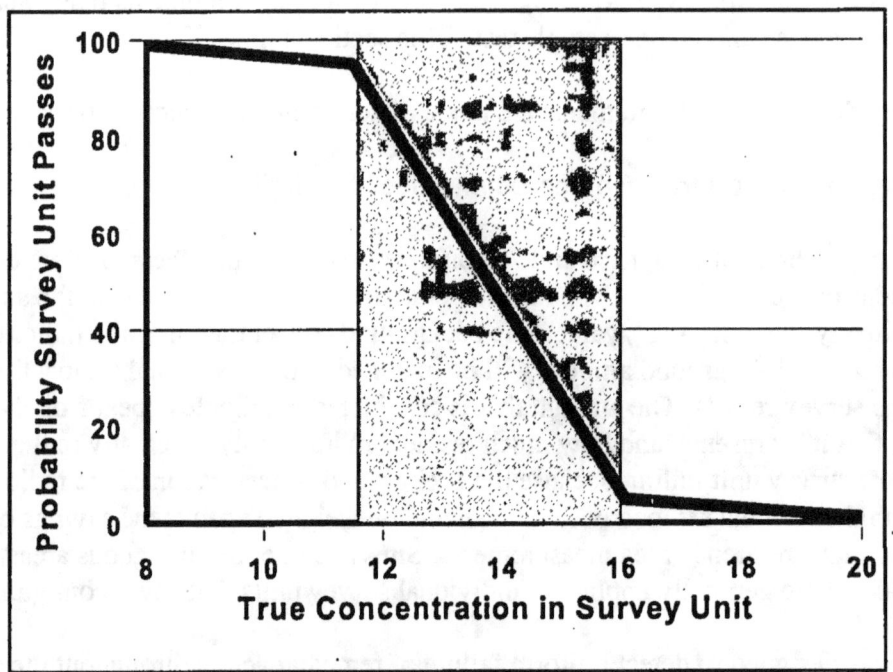

Figure 5.1 Desired Probability That the Survey Unit Passes

The data were taken on a triangular grid. The posting plot is shown in Figure 5.2. It is clear from this plot that there is residual radioactivity that is higher near the center of the survey unit and that diminishes as the survey unit boundary is approached. It is near zero on the west side.

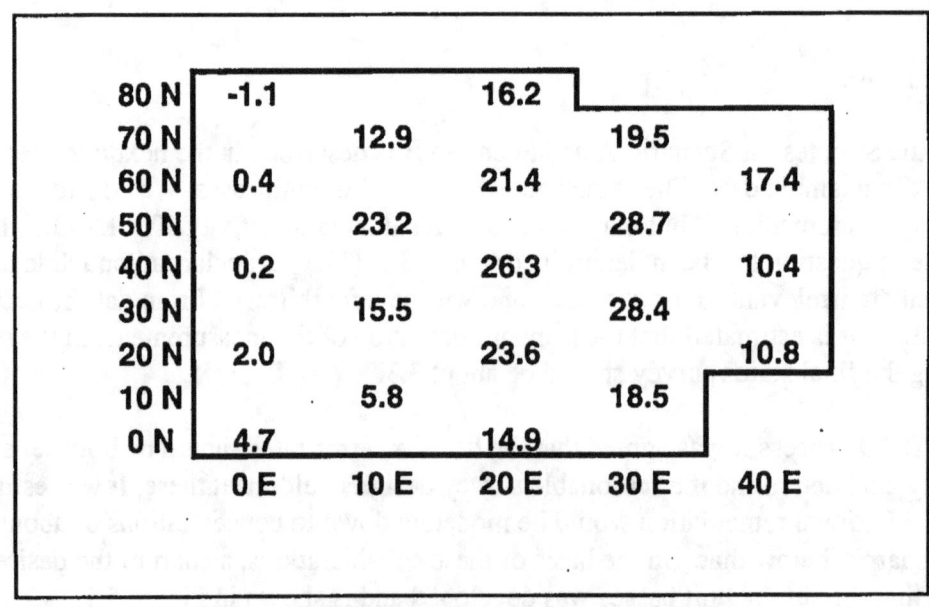

	0 E	10 E	20 E	30 E	40 E
80 N	-1.1		16.2		
70 N		12.9		19.5	
60 N	0.4		21.4		17.4
50 N		23.2		28.7	
40 N	0.2		26.3		10.4
30 N		15.5		28.4	
20 N	2.0		23.6		10.8
10 N		5.8		18.5	
0 N	4.7		14.9		

Figure 5.2 Posting Plot of Survey Unit Data

The one reported negative value stands out. Negative values can occur whenever an instrument, analysis or blank background is subtracted to obtain a net reading for a sample[1], which is then converted to a concentration.

Summary statistics for these data are shown in Table 5.1.

Table 5.1 Summary Statistics for Example Data of Figure 5.2

Mean	14.3
Standard Error	2.07
Median	15.5
Standard Deviation	9.5
Sample Variance	90.0
Kurtosis	−1.09
Skewness	−0.18
Range	29.8
Minimum	−1.1
Maximum	28.7
Count	21

The mean is 14.3, which in this case is actually less than the median, 15.5. This is the influence of the few low values on the west part of the survey unit, and is also the cause for the slight negative skewness. The range, 29.8, is only about three times the standard deviation, 9.5. This standard deviation, however, is three times larger than the value of 3.3 assumed in the survey design. If the null hypothesis is accepted, the effect of this higher standard deviation on the power of the test should be investigated. Recall that lower power in Scenario A means that more survey units with concentrations that are actually lower than the $DCGL_w$ are apt to fail the test. In Scenario B, lower power means that more survey units with concentrations greater than the $DCGL_w$ are apt to pass the test. A retrospective power curve, calculated as described in Chapter 10, can be used to decide if the error rates achieved are acceptable. A retrospective power analysis is not necessary when the null hypothesis is rejected, since the Type I error rate, α, is fixed at the design value when the critical value for the test is determined. However, taking additional samples to increase the power will increase the Type I error rate, unless provision for a two-stage test is made during the survey design. This is discussed further in Section 14.2.

A histogram of the data is shown in Figure 5.3. Except for the one negative value in the "zero" bin, the distribution appears reasonably symmetric.

---

[1] For samples with no residual radioactivity, negative values should occur about 50% of the time. It does not imply the existence of negative concentrations. Due to random fluctuations, the measurement merely happened to be lower during the sample analysis than it was during the background determination.

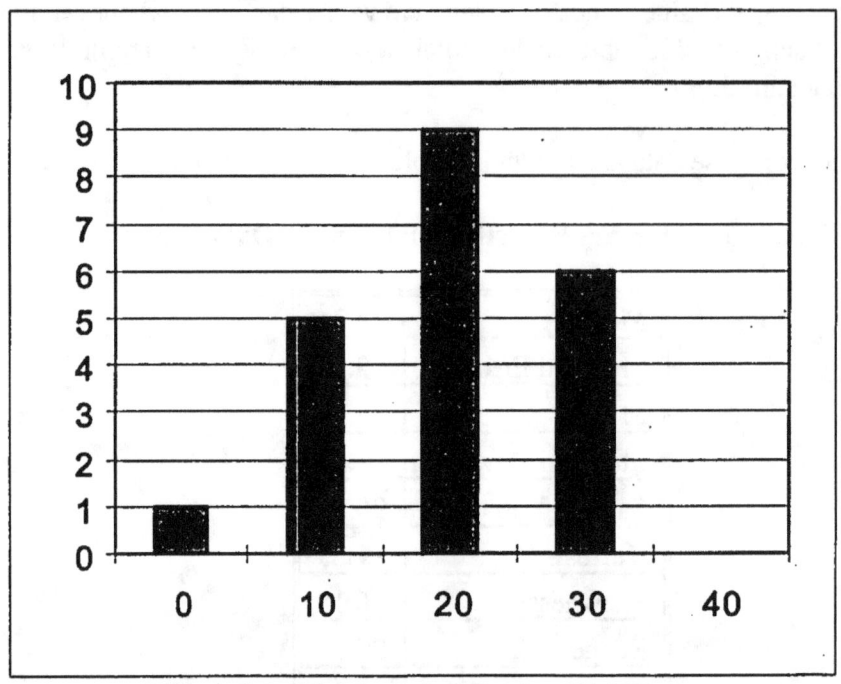

Figure 5.3 Histogram of Survey Unit Data

Further insight into the distribution of the data can be obtained using a ranked data plot. Table 5.2 shows the survey unit data ranked in increasing order. From this table it is already apparent that there is a large gap between the 6th and 7th ranked values. This is even clearer in the ranked data plot of Figure 5.4. It seems that there is a mixture of two distributions in the survey unit data. One is a distribution of low values near the western edge of the survey unit, and one of higher values in the rest of the survey unit. This partially explains the negative skewness and the large standard deviation.

Table 5.2 Ranked Data for Example of Figure 5.2

Rank	1	2	3	4	5	6	7
Measurement	-1.1	0.2	0.4	2	4.7	5.8	10.4
Rank	8	9	10	11	12	13	14
Measurement	10.8	12.9	14.9	15.5	16.2	17.4	18.5
Rank	15	16	17	18	19	20	21
Measurement	19.5	21.4	23.2	23.6	26.3	28.4	28.7

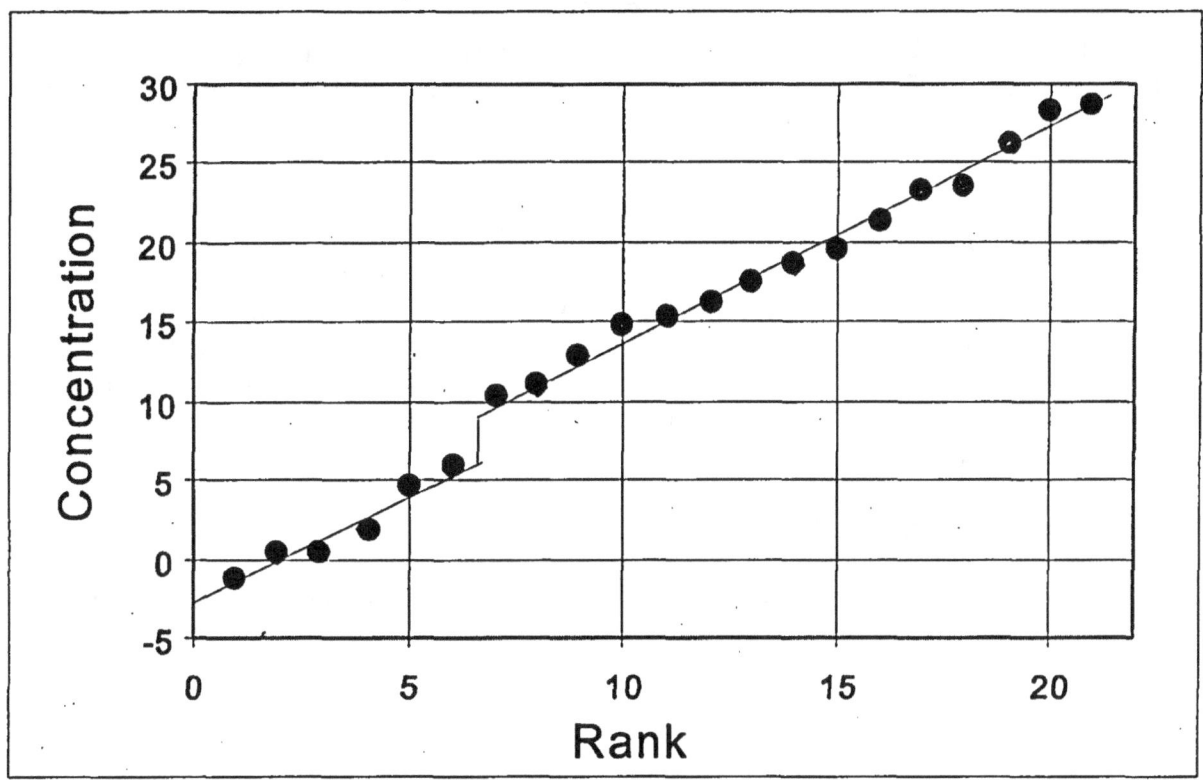

Figure 5.4 Ranked Data Plot of Survey Unit Data

5.2 Applying the Sign Test: Scenario A

The hypothesis tested by the Sign test under Scenario A is:

Null Hypothesis:

H_0: The median concentration of residual radioactivity in the survey unit is greater than the

   $DCGL_w$.

versus

Alternative Hypothesis:

H_a: The median concentration of residual radioactivity in the survey unit is less than the LBGR.

In order to use the one-sample Sign test, background concentrations of the radionuclide of concern are considered to be either zero or insignificant in comparison to the $DCGL_w$ Thus, there is no reference to background in statement of the null and alternative hypothesis. The null hypothesis is assumed to be true unless the statistical test indicates that it should be rejected in favor of the alternative. The parameter of interest is the mean concentration. The median is equal to the mean when the measurement distribution is symmetric, and is an approximation otherwise.

The null hypothesis states that the probability of a measurement less than the $DCGL_W$ is less than one-half, i.e., the 50th percentile (or median) is greater than the $DCGL_W$. The median is the concentration that would be exceeded by 50% of the measurements. Note that some individual survey unit measurements may exceed the $DCGL_W$ even when the survey unit as a whole meets the release criterion. In fact, a survey unit that averages close to the $DCGL_W$ might have almost half of its individual measurements greater than the $DCGL_W$. Such a survey unit may still meet the release criterion.

The hypothesis specifies a release criterion in terms of a $DCGL_W$ which is calculated as described in Section 3.3. The test should have sufficient power ($1-\beta$, as specified in the DQOs) to detect residual radioactivity concentrations at the Lower Boundary of the Gray Region (LBGR). If σ is the standard deviation of the measurements in the survey unit, then Δ/σ expresses the size of the shift (i.e., $\Delta = DCGL_W - LBGR$) as the number of standard deviations that would be considered large for the distribution of measurements in the survey unit. The procedure for determining Δ/σ was given in Section 3.8.1.

The Sign test is applied as follows in Scenario A:

(1) List the survey unit measurements, X_i, $i = 1, 2, 3..., N$. If a measurement is listed as "less than" a given value, insert that value for the measurement.

(2) Subtract each measurement, X_i, from the $DCGL_W$ to obtain the differences:
$D_i = DCGL_W - X_i$, $i = 1, 2, 3..., N$.

(3) If any difference is exactly zero, discard it from the analysis, and reduce the sample size, N, by the number of such zero measurements.

(4) Count the number of positive differences. The result is the test statistic S+. Note that a positive difference corresponds to a measurement below the $DCGL_W$ and contributes evidence that the survey unit meets the release criterion.

(5) Large values of S+ indicate that the null hypothesis is false. The value of S+ is compared to the table of critical values in Section A.3. If S+ is greater than the critical value, k, in that table, the null hypothesis is rejected.

For the example survey unit data, the calculations are shown in Table 5.3. Notice that when the data are ranked, it is really only necessary to observe the rank of the smallest measurement below the $DCGL_W$ in order to determine S+. However, the differences $D_i = DCGL_W - X_i = 15.9 - X_i$ are also shown. The number of positive differences, S+ = 11. The critical value of S+ for the Sign Test with $\alpha = 0.05$ and $N = 21$ is 14. Since S+ is less than 14, the null hypothesis cannot be rejected. The survey unit has failed the test.

Table 5.3 Calculations for Sign Test in Scenario A

Rank	Measurement	DCGL$_W$ – Measurement	Sign
1	−1.1	14.8	+
2	0.2	15.7	+
3	0.4	15.5	+
4	2.0	13.9	+
5	4.7	11.2	+
6	5.8	10.1	+
7	10.4	5.5	+
8	10.8	5.1	+
9	12.9	3.0	+
10	14.9	1.0	+
11	15.5	0.4	+
12	16.2	−0.3	−
13	17.4	−1.5	−
14	18.5	−2.6	−
15	19.5	−3.6	−
16	21.4	−5.5	−
17	23.2	−7.3	−
18	23.6	−7.7	−
19	26.3	−10.4	−
20	28.4	−12.5	−
21	28.7	−12.8	−

5.3 Applying the Sign Test: Scenario B

The hypothesis tested by the Sign test under Scenario B is:

Null Hypothesis:

H_0: The median concentration of residual radioactivity in the survey unit is less than the LBGR.

versus

Alternative Hypothesis:

H_a:The median concentration of residual radioactivity in the survey unit is greater than the DCGL$_W$.

In order to use the one-sample Sign test, background concentrations of the radionuclide of concern are considered to be either zero or insignificant in comparison to the LBGR. Thus, there is no reference to background in statement of the null and alternative hypothesis. The null hypothesis is assumed to be true unless the statistical test indicates that it should be rejected in favor of the alternative. The Type I error rate, α, is the probability that a survey unit with residual radioactivity at the LBGR will fail to be released. The power, $1-\beta$, is the probability that a survey unit with residual radioactivity at the $DCGL_W$ will fail to be released. The parameter of interest is the mean concentration. The median is equal to the mean when the measurement distribution is symmetric, and is an approximation otherwise.

The Sign test is carried out for Scenario B in a manner very similar to that for Scenario A:

(1) List the survey unit measurements, X_i, $i = 1, 2, 3..., N$. If a measurement is listed as "less than" a given value, insert that value for the measurement.

(2) Subtract the LBGR from each measurement, X_i, to obtain the differences:
   $D_i = X_i - LBGR$, $i = 1, 2, 3..., N$.

(3) If any difference is exactly zero, discard it from the analysis, and reduce the sample size, N, by the number of such zero measurements.

(4) Count the number of positive differences. The result is the test statistic S+. A positive difference corresponds to a measurement above the LBGR and is evidence that the median concentration of residual radioactivity survey unit may exceed it.

(5) Large values of S+ indicate that the null hypothesis is false. The value of S+ is compared to the table of critical values in Section A.3. If S+ is greater than the critical value, k, in that table, the null hypothesis is rejected.

For the example survey unit data, the calculations are shown in Table 5.4. Notice that when the data are ranked, it is really only necessary to observe the rank of the smallest measurement below the LBGR in order to determine S- $= N - $ S+. However, the differences $D_i = X_i - LBGR = X_i - 11.5$ are also shown. The number of positive differences, S+ $= 13$. The critical value of S+ for the Sign Test with $\alpha = 0.05$ and $N = 21$ is 14. Since S+ is less than 14, the null hypothesis cannot be rejected. The survey unit has passed the test. However, it remains to determine the power of the test. Since the observed standard deviation is much greater than that estimated for the test design, it is likely that this survey unit passed simply because there was insufficient power to detect residual radioactivity at the $DCGL_W$. The power calculation for this example is given in Section 10.1.

5.4 Interpretation of Test Results

Once the results of the statistical tests are obtained, the specific steps required to achieve site release will depend on the procedures described in the regulatory guide. The following are suggested considerations for the interpretation of the test results with respect to the release limit established for the site.

Table 5.4 Calculations for Sign Test in Scenario B

Rank	Measurement	Measurement - LBGR	Sign
1	−1.1	−12.6	−
2	0.2	−11.3	−
3	0.4	−11.1	−
4	2.0	−9.5	−
5	4.7	−6.8	−
6	5.8	−5.7	−
7	10.4	−1.1	−
8	10.8	−0.7	−
9	12.9	1.4	+
10	14.9	3.4	+
11	15.5	4.0	+
12	16.2	4.7	+
13	17.4	5.9	+
14	18.5	7.0	+
15	19.5	8.0	+
16	21.4	9.9	+
17	23.2	11.7	+
18	23.6	12.1	+
19	26.3	14.8	+
20	28.4	16.9	+
21	28.7	17.2	+

5.4.1 If the Null Hypothesis Is Not Rejected

Whenever the null hypothesis is not rejected, it is important to complete the analysis by performing a retrospective power analysis for the test. In Scenario A, this will ensure that further remediation is not required simply because the final status survey was not sensitive enough to detect residual radioactivity below the LBGR. In Scenario B, this will ensure that a survey unit is not released simply because the final status survey was not sensitive enough to detect residual radioactivity above the guideline level. The power analysis may be performed as indicated in Chapter 10, using the actual values of the number of measurements, N, and their observed measurement standard deviation s in place of σ. In some cases, a site specific simulation of the retrospective power may be warranted when sufficient power cannot be demonstrated by any of the other suggested methods.

5.4.2 If the Null Hypothesis Is Rejected

If the null hypothesis for the Sign test is rejected in Scenario A, it indicates that the residual radioactivity in the survey unit is less than the $DCGL_w$. However, it may still be necessary to document the concentration of residual radioactivity. It is generally best to use the average measured concentration for this purpose.

If the null hypothesis is rejected in Scenario B it indicates that the residual radioactivity in the survey unit exceeds the LBGR. In this case it is important to determine not only the average concentration of residual radioactivity in the survey unit, δ, but also whether this amount exceeds the release criteria. When the data are normally distributed, the average concentration is generally the best estimator for δ. However, when the data are not normally distributed, other estimators are often better for the same reasons that nonparametric tests are often better than the corresponding parametric tests. These methods are discussed by Lehmann and D'Abrera (1975). When the estimate for δ is below $DCGL_w$, the survey unit may be judged sufficiently remediated, subject to ALARA considerations. Otherwise, further remediation will generally be required.

The treatment of data that are below the limit of detection will be an important component of these calculations. Whenever possible, the actual results of the measurement should be reported, with an associated total uncertainty that includes both random and systematic errors. Replacing values below the MDC with the MDC value will cause δ to be overestimated.

6 WILCOXON RANK SUM TEST: CONTAMINANT PRESENT IN BACKGROUND

The statistical tests discussed in this section will be used to compare each survey unit with an appropriately chosen, site-specific reference area. Each reference area should be chosen on the basis of its similarity to the survey unit, as discussed in Section 2.2.7.

In Scenario A, the comparison of measurements from the reference area and survey unit is made using the Wilcoxon Rank Sum (WRS) test (also called the Mann-Whitney test).

Under Scenario B, the comparison of measurements in the reference area and survey unit is made using two nonparametric statistical tests: the WRS test and the Quantile test. The WRS and Quantile tests are both used because each test detects different residual contamination patterns in the survey units. Because two tests are used, the Type I error rate, α, specified during the DQO process is halved for the individual tests. The Quantile test is discussed in Chapter 7.

In addition to the statistical tests, the EMC is performed against each measurement to assure that it does not exceed a specified investigation level. If any measurement in the remediated survey unit exceeds the specified investigation level, then additional investigation is recommended, at least locally, regardless of the outcome of the WRS or Quantile test.

The WRS test is most effective when residual radioactivity is uniformly present throughout a survey unit. The test is designed to detect whether or not this activity exceeds the $DCGL_w$. The advantage of the nonparametric WRS test is that it does not assume that the data are normally or log-normally distributed. The WRS test also allows for less than detectable measurements in either the reference area or the survey unit. As a general rule, the WRS test can be used with up to 40% less than detectable measurements in the reference area and the survey unit combined. However, the use of less than values in data reporting is not encouraged. Wherever possible, the actual result of a measurement, together with its uncertainty, should be reported.

6.1 Introduction

The use of the WRS test in Scenario A and Scenario B are described in the next two sections, illustrated with example data. We consider a Class 2 survey consisting of interior drywall surfaces, that may have some residual radioactivity. The $DCGL_w$ for the radionuclide in question has been determined to be 160. (The particular radionuclide and units of measurement are irrelevant to the example, and will be left arbitrary.) The background level is about 40. It is estimated that the standard deviation of the measurements in the survey unit and the reference area is about 6.

Since the $DCGL_w$ is so much larger than σ, large sample sizes will not be needed even if the acceptable error rates are set to low values. In this circumstance, the rule of thumb that Δ/σ should lie between one and three can be used to set the Lower Bound of the Gray Region (LBGR). If, for example, $\Delta/\sigma = 3$, then $\Delta = 18$, since σ is estimated at 6. The Lower Bound of the Gray Region is then LBGR = $DCGL_w - \Delta = 160 - 18 = 142$. If the decision error rates are both set equal to 0.05 initially, then from Table 3.3, for $\alpha = \beta = 0.05$ with $\Delta/\sigma = 3$, ten samples each are required in the reference area and the survey unit.

Under Scenario B, the Type I error rate is halved, so $\alpha_w = \alpha/2 = 0.05/2 = 0.025$. Then, again from Table 3.3, for $\alpha = 0.025$, $\beta = 0.05$, and $\Delta/\sigma = 3$, twelve samples each are required in the reference area and the survey unit. The corresponding chart of the desired probability that the survey unit passes is shown in Figure 6.1. Note that, although the probability that a survey unit at the LBGR passes the WRS test in Scenario B is 97.5%, the overall probability of passing both the WRS and Quantile tests is approximately 95%.

Since both scenarios are illustrated using the same set of data, the larger sample size will be used for both.

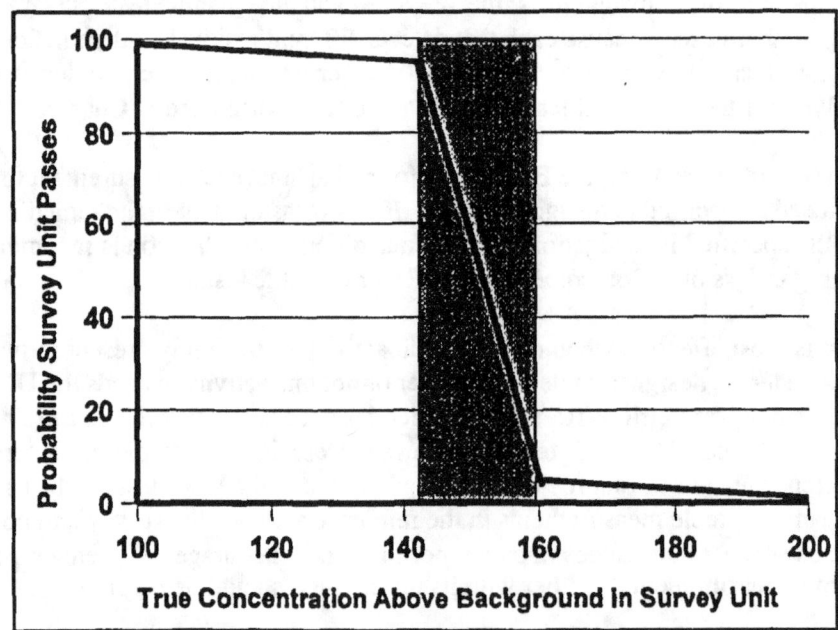

Figure 6.1 Desired Probability That the Survey Unit Passes

The data were taken on a triangular grid[1], and the posting plot is shown in Figure 6.2. For this example the concentration of the radionuclide of interest is given in arbitrary units. It is clear from this plot that there is residual radioactivity above background in the survey unit.

Summary statistics for these data are shown in Table 6.1. The mean and median are fairly close in both the reference area and the survey unit. The standard deviations of the data are slightly larger than estimated during the survey design, but the ratio Δ/σ remains above 2, so the impact on the power of the tests should not be severe. The range of the data is between 3 and 4 standard

---

[1]A random start systematic grid is used in Class 2 and 3 survey units primarily to limit the size of any potential elevated areas. Since areas of elevated activity are not an issue in the reference areas, the measurement locations can be either random or on a random start systematic grid.

deviations, which is about right for these sample sizes (see Figure 4.1).

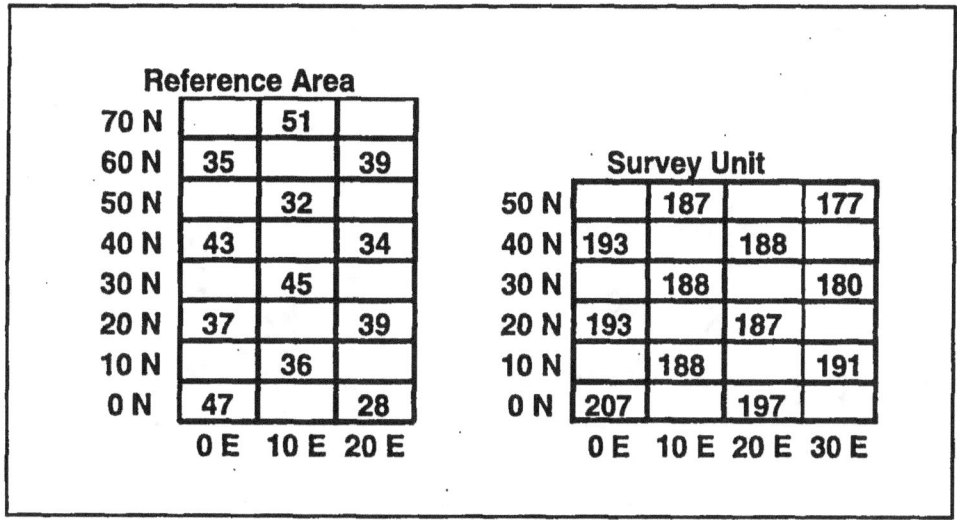

Figure 6.2 Posting Plot of Reference Area and Survey Unit Data

Table 6.1 Summary Statistics for Example Data of Figure 6.2

Reference Area			Survey Unit	
Mean	38.8		Mean	189.8
Median	38		Median	188
Std Dev	6.6		Std Dev	8.1
Kurtosis	−0.4		Kurtosis	2.2
Skewness	0.3		Skewness	0.9
Range	23		Range	32
Minimum	28		Minimum	177
Maximum	51		Maximum	209
Count	12		Count	12

A histogram of the data is shown in Figure 6.3. The data distributions are fairly symmetric. The survey unit and reference area distributions are clearly separated by an amount much larger than the width of either. The difference in the medians is 188 − 38 = 150, and the difference in the means is 189.8 − 38.8 = 151. Both of these values are very close to the $DCGL_w$ of 160. It is in just such cases that the statistical tests are most useful in determining the significance of these values.

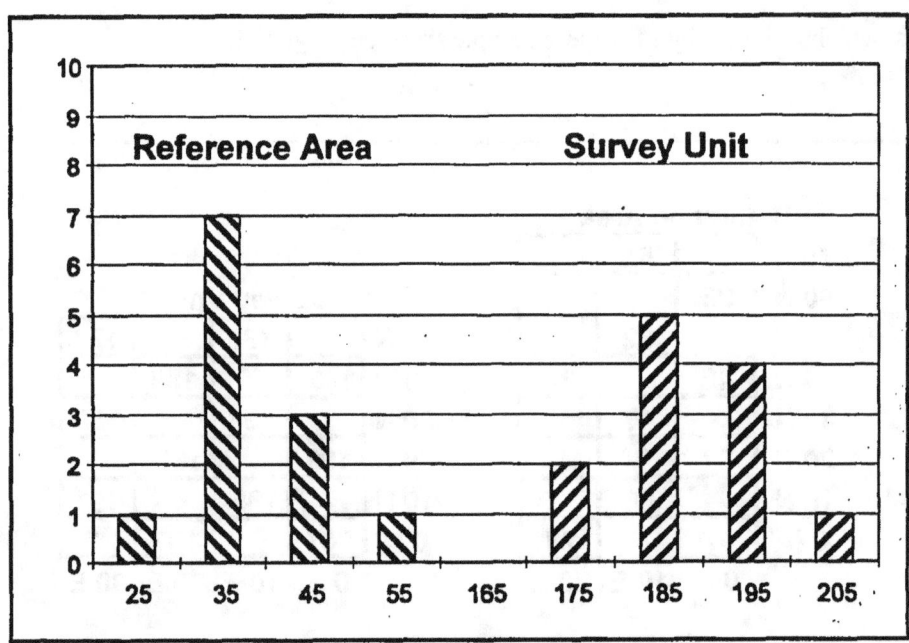

Figure 6.3 Histograms of Reference Area and Survey Unit Data

Since there are equal numbers of data in the reference area and in the survey unit, a Quantile-Quantile plot is easily constructed. Table 6.2 shows the reference area and survey unit data each separately ranked in increasing order. The pairs of data from the reference area and the survey unit with the same rank are plotted in Figure 6.4. This is the Quantile-Quantile plot. The position of the medians is indicated by the solid bar and the central 50% of the data is enclosed in the dashed box. The plot is fairly straight, and the slope is not greatly different from one, indicating the the shapes of the reference area and survey unit distributions are similar. However, the survey unit distribution is shifted to values about 150 larger. Again, the significance of this relative to the $DCGL_w$ is precisely what the WRS test is designed to determine.

Table 6.2 Ranked Data for Example of Figure 6.2

Rank	Reference Area	Rank	Survey Unit
1	28	1	177
2	32	2	180
3	34	3	187
4	35	4	187
5	36	5	188
6	37	6	188
7	39	7	188
8	39	8	191
9	43	9	193
10	45	10	193
11	47	11	197
12	51	12	207

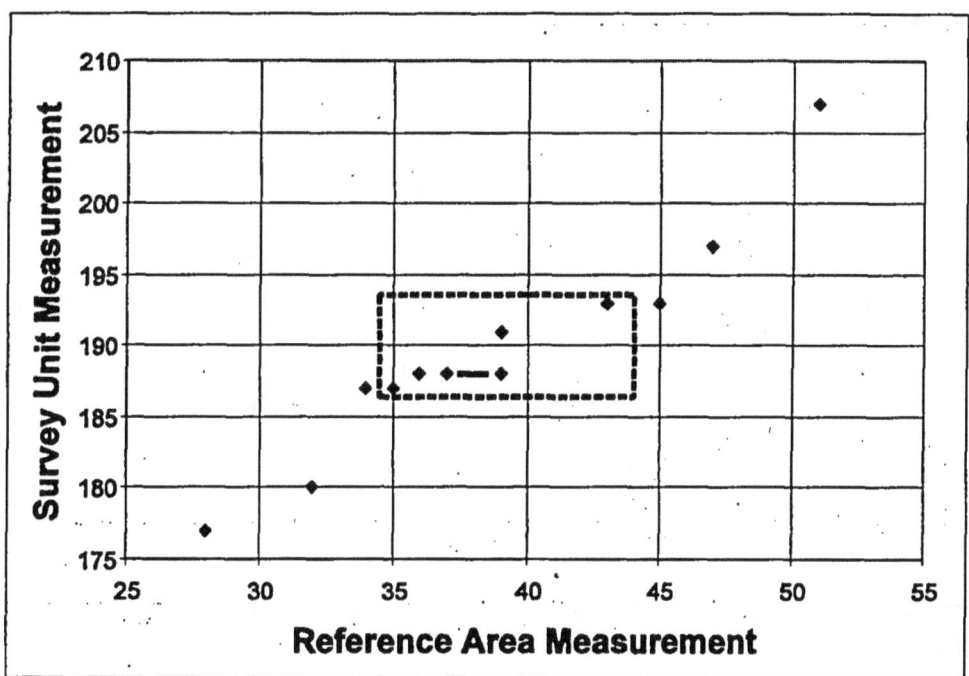

Figure 6.4 Quantile-Quantile Plot of Example Data

6.2 Applying the WRS Test: Scenario A

The hypothesis tested by the WRS test under Scenario A is:

Null Hypothesis:

H_0: The median concentration in the survey unit exceeds that in the reference area by more than

the $DCGL_W$.

versus

Alternative Hypothesis:

H_a: The median concentration in the survey unit exceeds that in the reference area by less than

the LBGR.

The null hypothesis is assumed to be true unless the statistical test indicates that it should be rejected in favor of the alternative. One assumes that any difference between the reference area and survey unit concentration distributions is due to a shift in the survey unit concentrations to higher values—i.e., due to the presence of residual radioactivity in addition to background. The size of this shift is the difference in the mean concentrations. The median is equal to the mean when the measurement distributions are symmetric, and is an approximation otherwise. Note that some or all of the survey unit measurements may be larger than some reference area measurements, while still meeting the release criterion. Indeed, some survey unit measurements may exceed some reference area measurements by more than the $DCGL_w$. The result of the hypothesis test determines whether or not the survey unit as a whole is deemed to meet the release criterion. The EMC is used to screen individual measurements.

Assumptions underlying this test are that (1) the samples from the reference area and the survey unit are independent random samples, and (2) each measurement is independent of every other measurement—regardless of the set of samples from which it came.

The hypothesis specifies a release criterion in terms of a $DCGL_w$ which is calculated as described in Section 3.3. The test should have sufficient power $(1-\beta$, as specified in the DQOs) to detect residual radioactivity concentrations at the Lower Boundary of the Gray Region (LBGR). If σ is the standard deviation of the measurements in the survey unit, then Δ/σ expresses the size of the shift (i.e. $\Delta = DCGL_w - LBGR$) as the number of standard deviations that would be considered large for the distribution of measurements in the survey unit. The procedure for determining Δ/σ was given in Section 3.8.1.

The WRS test is applied as follows under Scenario A:

(1) Obtain the adjusted reference area measurements, Z_i, by adding the $DCGL_w$ to each reference area measurement, X_i. $Z_i = X_i + DCGL_w$

(2) The m adjusted reference area sample measurements, Z_i, and the n survey unit sample measurements, Y_i, are pooled and ranked in order of increasing size from 1 to N, where $N = m+n$.

(3) If several measurements are tied (have the same value), they are all assigned the average rank of that group of tied measurements.

(4) If there are t less than detectable values, they are all assigned the rank $(t + 1)/2$. If there is more than one detection limit, all observations below the largest detection limit should be treated as less than detected. If more than 40% of the data from either the reference area or survey unit are less than detectable, the WRS test *cannot* be used. As stated previously, the use of less than values in data reporting is not encouraged. Wherever possible, the actual result of a measurement, together with its uncertainty, should be reported.

(5) Sum the ranks of the adjusted measurements from the reference area, W_r. Note that since the sum of the first N integers is $N(N+1)/2$, one can equivalently sum the ranks of the measurements from the survey unit, W_s, and compute $W_r = N(N + 1)/2 - W_s$.

(6) Compare W_r with the critical value given in Table A.4 for the appropriate values of n, m, and α. If W_r is greater than the tabulated value, reject the hypothesis that the survey unit exceeds the release criterion.

The data for the example are shown in column A of Table 6.3 In column B, the code R was inserted to denote a reference area measurement, and S to denote a survey unit measurement. In column A, the data are simply listed as they were obtained. Column C contains the adjusted data. The adjusted data are obtained by adding the $DCGL_w$ to the reference area measurements.

The ranks of the adjusted data appear in Column D. They range from 1 to 24, since there is a total of 12 + 12 measurements. The sum of all the ranks is $N(N + 1)/2 = (24)(25)/2 = 300$. Column E contains only the ranks belonging to the adjusted reference area measurements. The sum of the ranks of the adjusted reference area data is 199. From Table A.4, for $\alpha = \beta = 0.05$ and $n = m = 12$, we find that the critical value is 179. Thus, the sum of the reference area ranks, 199, is greater than the critical value, 179, and the null hypothesis that the survey unit concentrations exceed the DCGL$_W$ is rejected. In Scenario A, the survey unit passes.

The analysis for the WRS test is very well suited to the use of a computer spreadsheet. The spreadsheet formulas in Microsoft Excel™ (1993) used for the example above are given in Table 6.4.

Table 6.3 WRS Test for Class 2 Interior Drywall Survey Unit
(Measurements from the reference area and the survey unit are denoted by R and S, respectively)

	A	B	C	D	E
1	Data	Area	Adjusted Data	Ranks	Reference Area Ranks
2	47	R	207	22	22
3	28	R	188	6.5	6.5
4	36	R	196	15	15
5	37	R	197	16.5	16.5
6	39	R	199	18.5	18.5
7	45	R	205	21	21
8	43	R	203	20	20
9	34	R	194	13	13
10	32	R	192	10	10
11	35	R	195	14	14
12	39	R	199	18.5	18.5
13	51	R	211	24	24
14	209	S	209	23	—
15	197	S	197	16.5	—
16	188	S	188	6.5	—
17	191	S	191	9	—
18	193	S	193	11.5	—
19	187	S	187	3.5	—
20	188	S	188	6.5	—
21	180	S	180	2	—
22	193	S	193	11.5	—
23	188	S	188	6.5	—
24	187	S	187	3.5	—
25	177	S	177	1	—
26			Sum =	300	199

Table 6.4 Spreadsheet Formulas Used in Table 6.3

	A	B	C	D	E
1	Data	Area	**Adjusted Data**	**Ranks**	**Reference Ranks**
2	47	R	=IF(B2="R",A2+160,A2)	=RANK(C2,$C$2:$C$25,1)+(COUNTIF($C$2:$C$25,C2) - 1) / 2	=IF(B2="R",D2,0)
3	28	R	=IF(B3="R",A3+160,A3)	=RANK(C3,$C$2:$C$25,1)+(COUNTIF($C$2:$C$25,C3) - 1) / 2	=IF(B3="R",D3,0)
4	36	R	=IF(B4="R",A4+160,A4)	=RANK(C4,$C$2:$C$25,1)+(COUNTIF($C$2:$C$25,C4) - 1) / 2	=IF(B4="R",D4,0)
5	37	R	=IF(B5="R",A5+160,A5)	=RANK(C5,$C$2:$C$25,1)+(COUNTIF($C$2:$C$25,C5) - 1) / 2	=IF(B5="R",D5,0)
6	39	R	=IF(B6="R",A6+160,A6)	=RANK(C6,$C$2:$C$25,1)+(COUNTIF($C$2:$C$25,C6) - 1) / 2	=IF(B6="R",D6,0)
7	45	R	=IF(B7="R",A7+160,A7)	=RANK(C7,$C$2:$C$25,1)+(COUNTIF($C$2:$C$25,C7) - 1) / 2	=IF(B7="R",D7,0)
8	43	R	=IF(B8="R",A8+160,A8)	=RANK(C8,$C$2:$C$25,1)+(COUNTIF($C$2:$C$25,C8) - 1) / 2	=IF(B8="R",D8,0)
9	34	R	=IF(B9="R",A9+160,A9)	=RANK(C9,$C$2:$C$25,1)+(COUNTIF($C$2:$C$25,C9) - 1) / 2	=IF(B9="R",D9,0)
10	32	R	=IF(B10="R",A10+160,A10)	=RANK(C10,$C$2:$C$25,1)+(COUNTIF($C$2:$C$25,C10) - 1) / 2	=IF(B10="R",D10,0)
11	35	R	=IF(B11="R",A11+160,A11)	=RANK(C11,$C$2:$C$25,1)+(COUNTIF($C$2:$C$25,C11) - 1) / 2	=IF(B11="R",D11,0)
12	39	R	=IF(B12="R",A12+160,A12)	=RANK(C12,$C$2:$C$25,1)+(COUNTIF($C$2:$C$25,C12) - 1) / 2	=IF(B12="R",D12,0)
13	51	R	=IF(B13="R",A13+160,A13)	=RANK(C13,$C$2:$C$25,1)+(COUNTIF($C$2:$C$25,C13) - 1) / 2	=IF(B13="R",D13,0)
14	209	S	=IF(B14="R",A14+160,A14)	=RANK(C14,$C$2:$C$25,1)+(COUNTIF($C$2:$C$25,C14) - 1) / 2	=IF(B14="R",D14,0)
15	197	S	=IF(B15="R",A15+160,A15)	=RANK(C15,$C$2:$C$25,1)+(COUNTIF($C$2:$C$25,C15) - 1) / 2	=IF(B15="R",D15,0)
16	188	S	=IF(B16="R",A16+160,A16)	=RANK(C16,$C$2:$C$25,1)+(COUNTIF($C$2:$C$25,C16) - 1) / 2	=IF(B16="R",D16,0)
17	191	S	=IF(B17="R",A17+160,A17)	=RANK(C17,$C$2:$C$25,1)+(COUNTIF($C$2:$C$25,C17) - 1) / 2	=IF(B17="R",D17,0)
18	193	S	=IF(B18="R",A18+160,A18)	=RANK(C18,$C$2:$C$25,1)+(COUNTIF($C$2:$C$25,C18) - 1) / 2	=IF(B18="R",D18,0)
19	187	S	=IF(B19="R",A19+160,A19)	=RANK(C19,$C$2:$C$25,1)+(COUNTIF($C$2:$C$25,C19) - 1) / 2	=IF(B19="R",D19,0)
20	188	S	=IF(B20="R",A20+160,A20)	=RANK(C20,$C$2:$C$25,1)+(COUNTIF($C$2:$C$25,C20) - 1) / 2	=IF(B20="R",D20,0)
21	180	S	=IF(B21="R",A21+160,A21)	=RANK(C21,$C$2:$C$25,1)+(COUNTIF($C$2:$C$25,C21) - 1) / 2	=IF(B21="R",D21,0)
22	193	S	=IF(B22="R",A22+160,A22)	=RANK(C22,$C$2:$C$25,1)+(COUNTIF($C$2:$C$25,C22) - 1) / 2	=IF(B22="R",D22,0)
23	188	S	=IF(B23="R",A23+160,A23)	=RANK(C23,$C$2:$C$25,1)+(COUNTIF($C$2:$C$25,C23) - 1) / 2	=IF(B23="R",D23,0)
24	187	S	=IF(B24="R",A24+160,A24)	=RANK(C24,$C$2:$C$25,1)+(COUNTIF($C$2:$C$25,C24) - 1) / 2	=IF(B24="R",D24,0)
25	177	S	=IF(B25="R",A25+160,A25)	=RANK(C25,$C$2:$C$25,1)+(COUNTIF($C$2:$C$25,C25) - 1) / 2	=IF(B25="R",D25,0)
26			Sum =	299	=SUM(E2:E25)

Note that some spreadsheet programs assign the *lowest* rank in a group of ties to every member of the group, rather than the *average* rank for the group. This can be corrected by adding to each rank the value $(g - 1)/2$, where g is the number of data points in the group. This is the function of the COUNTIF statement in column D of Table 6.4.

6.3 Applying the WRS Test: Scenario B

Two tests are used in Scenario B to ensure that there is adequate power to detect a survey unit that does not meet the release criterion. The concept of the statistical power of a test was discussed in Section 2.3.2. The WRS test has more power than the Quantile test to detect uniform failure of remedial action throughout the survey unit. The Quantile test has more power than the WRS test to detect failure of remedial action in only a few areas within the survey unit. These nonparametric tests do not require that the data be normally or log-normally distributed. Measurements reported as non-detects may also be used with these tests, although this practice is discouraged[2]. As a general rule, the WRS test can be used with up to 40% less than detectable measurements in either the reference area or the survey unit. The Quantile test can be used even when more than 50% of the measurements are below the limit of detection.

In addition, an elevated measurement comparison is conducted. This consists of determining if any measurements in the remediated survey unit exceed a specified investigation level. If so, then additional investigation is required, at least locally, regardless of the outcome of the WRS and Quantile tests.

The hypothesis tested by the WRS test under Scenario B is:

Null Hypothesis:

H_0: The difference in the median concentration of radioactivity in the survey unit and in the reference area is less than the LBGR.

versus

Alternative Hypothesis:

H_a: The difference in the median concentration of radioactivity in the survey unit and in the reference area is greater than the $DCGL_w$.

The Type I error rate, $\alpha_w = \alpha/2$, is the probability that a survey unit with residual radioactivity (above background) at the LBGR will fail this test. The power, $1-\beta$, is the probability that a survey unit with residual radioactivity at the $DCGL_w$ will fail this test.

The WRS test is applied as follows under Scenario B:

(1) Obtain the adjusted survey unit measurements, Z_i, by subtracting the LBGR from each

---

[2] All actual measurement results (with an associated uncertainty) should be reported, even if they are negative, so that unbiased estimates of averages can be calculated.

survey unit measurement, Y_i. $Z_i = Y_i - LBGR$

(2) The n adjusted survey unit measurements, Z_i, and the m reference area measurements, X_i, are pooled and ranked in order of increasing size from 1 to N, where $N = m+n$.

(3) If several measurements are tied (have the same value), they are all assigned the average rank of that group of tied measurements.

(4) If there are t less than detectable values, they are all assigned the rank $(t + 1)/2$. If there is more than one detection limit, all observations below the largest detection limit should be treated as less than detectable. If more than 40% of the data from either the reference area or survey unit are less than detectable, the WRS test *cannot* be used.

(5) Sum the ranks of the adjusted measurements from the survey unit, W_s. Note that since the sum of the first N integers is $N(N + 1)/2$, one can equivalently sum the ranks of the measurements from the reference area, W_r, and compute $W_s = N(N + 1)/2 - W_r$.

(6) Compare W_s with the critical value given in Table A.4 for the appropriate values of n, m, and α. If W_s is greater than the tabulated value, reject the hypothesis that the difference in the median concentration between the survey unit and the reference area is less than the LBGR.

The data for the example are shown in column A of Table 6.5 In column B, the code R was inserted to denote a reference area measurement, and S to denote a survey unit measurement. In column A, the data are simply listed as they were obtained. Column C contains the adjusted data. The adjusted data are obtained by subtracting the LBGR from the survey unit measurements.

The ranks of the adjusted data appear in Column D. They range from 1 to 24, since there is a total of 12 + 12 measurements. The sum of all the ranks is $N(N + 1)/2 = (24)(25)/2 = 300$. Column E contains only the ranks belonging to the adjusted survey unit measurements. The sum of the ranks of the adjusted survey unit data is 194.5. From Table A.4, for $\alpha_w = \alpha/2 = 0.025$, $\beta = 0.05$, and $n = m = 12$, we find that the critical value is 184. Thus, the sum of the adjusted survey unit ranks, 194.5, is greater than the critical value, 184, and the null hypothesis that the survey unit concentrations do not exceed LBGR is rejected. In Scenario B, the true survey unit residual radioactivity is judged to be in excess of 142 above background.

The analysis for the WRS test is very well suited to the use of a computer spreadsheet. The spreadsheet formulas in Microsoft Excel™ (1993) used for the example above are given in Table 6.6. Note that some spreadsheet programs assign the *lowest* rank in a group of ties to every member of the group, rather than the *average* rank for the group. This can be corrected by adding to each rank the value $(g - 1)/2$, where g is the number of data points in the group. This is the function of the COUNTIF statement in column D of Table 6.6.

Table 6.5 WRS Test Under Scenario B for Class 2 Interior Drywall Survey Unit
(Measurements from the reference area and the survey unit are denoted by R and S, respectively)

	A	B	C	D	E
1	Data	Area	Adjusted Data	Ranks	Survey Unit Ranks
2	47	R	47	18	—
3	28	R	28	1	—
4	36	R	36	6	—
5	37	R	37	7	—
6	39	R	39	9.5	—
7	45	R	45	13	—
8	43	R	43	11	—
9	34	R	34	3	—
10	32	R	32	2	—
11	35	R	35	4.5	—
12	39	R	39	9.5	—
13	51	R	51	21	—
14	209	S	67	24	24
15	197	S	55	23	23
16	188	S	46	16	16
17	191	S	49	19	19
18	193	S	51	21	21
19	187	S	45	13	13
20	188	S	46	16	16
21	180	S	38	8	8
22	193	S	51	21	21
23	188	S	46	16	16
24	187	S	45	13	13
25	177	S	35	4.5	4.5
26			Sum =	300	194.5

Table 6.6 Spreadsheet Formulas Used in Table 6.5

	A	B	C	D	E
1	Data	Area	Adjusted Data	Ranks	Survey Unit Ranks
2	47	R	=IF(B2="S",A2-142,A2)	=RANK(C2,$C$2:$C$25,1) +(COUNTIF($C$2:$C$25,C2) - 1) / 2	=IF(B2="S",D2,0)
3	28	R	=IF(B3="S",A3-142,A3)	=RANK(C3,$C$2:$C$25,1) +(COUNTIF($C$2:$C$25,C3) - 1) / 2	=IF(B3="S",D3,0)
4	36	R	=IF(B4="S",A4-142,A4)	=RANK(C4,$C$2:$C$25,1) +(COUNTIF($C$2:$C$25,C4) - 1) / 2	=IF(B4="S",D4,0)
5	37	R	=IF(B5="S",A5-142,A5)	=RANK(C5,$C$2:$C$25,1) +(COUNTIF($C$2:$C$25,C5) - 1) / 2	=IF(B5="S",D5,0)
6	39	R	=IF(B6="S",A6-142,A6)	=RANK(C6,$C$2:$C$25,1) +(COUNTIF($C$2:$C$25,C6) - 1) / 2	=IF(B6="S",D6,0)
7	45	R	=IF(B7="S",A7-142,A7)	=RANK(C7,$C$2:$C$25,1) +(COUNTIF($C$2:$C$25,C7) - 1) / 2	=IF(B7="S",D7,0)
8	43	R	=IF(B8="S",A8-142,A8)	=RANK(C8,$C$2:$C$25,1) +(COUNTIF($C$2:$C$25,C8) - 1) / 2	=IF(B8="S",D8,0)
9	34	R	=IF(B9="S",A9-142,A9)	=RANK(C9,$C$2:$C$25,1) +(COUNTIF($C$2:$C$25,C9) - 1) / 2	=IF(B9="S",D9,0)
10	32	R	=IF(B10="S",A10-142,A10)	=RANK(C10,$C$2:$C$25,1) +(COUNTIF($C$2:$C$25,C10) - 1) / 2	=IF(B10="S",D10,0)
11	35	R	=IF(B11="S",A11-142,A11)	=RANK(C11,$C$2:$C$25,1) +(COUNTIF($C$2:$C$25,C11) - 1) / 2	=IF(B11="S",D11,0)
12	39	R	=IF(B12="S",A12-142,A12)	=RANK(C12,$C$2:$C$25,1) +(COUNTIF($C$2:$C$25,C12) - 1) / 2	=IF(B12="S",D12,0)
13	51	R	=IF(B13="S",A13-142,A13)	=RANK(C13,$C$2:$C$25,1) +(COUNTIF($C$2:$C$25,C13) - 1) / 2	=IF(B13="S",D13,0)
14	209	S	=IF(B14="S",A14-142,A14)	=RANK(C14,$C$2:$C$25,1) +(COUNTIF($C$2:$C$25,C14) - 1) / 2	=IF(B14="S",D14,0)
15	197	S	=IF(B15="S",A15-142,A15)	=RANK(C15,$C$2:$C$25,1) +(COUNTIF($C$2:$C$25,C15) - 1) / 2	=IF(B15="S",D15,0)
16	188	S	=IF(B16="S",A16-142,A16)	=RANK(C16,$C$2:$C$25,1) +(COUNTIF($C$2:$C$25,C16) - 1) / 2	=IF(B16="S",D16,0)
17	191	S	=IF(B17="S",A17-142,A17)	=RANK(C17,$C$2:$C$25,1) +(COUNTIF($C$2:$C$25,C17) - 1) / 2	=IF(B17="S",D17,0)
18	193	S	=IF(B18="S",A18-142,A18)	=RANK(C18,$C$2:$C$25,1) +(COUNTIF($C$2:$C$25,C18) - 1) / 2	=IF(B18="S",D18,0)
19	187	S	=IF(B19="S",A19-142,A19)	=RANK(C19,$C$2:$C$25,1) +(COUNTIF($C$2:$C$25,C19) - 1) / 2	=IF(B19="S",D19,0)
20	188	S	=IF(B20="S",A20-142,A20)	=RANK(C20,$C$2:$C$25,1) +(COUNTIF($C$2:$C$25,C20) - 1) / 2	=IF(B20="S",D20,0)
21	180	S	=IF(B21="S",A21-142,A21)	=RANK(C21,$C$2:$C$25,1) +(COUNTIF($C$2:$C$25,C21) - 1) / 2	=IF(B21="S",D21,0)
22	193	S	=IF(B22="S",A22-142,A22)	=RANK(C22,$C$2:$C$25,1) +(COUNTIF($C$2:$C$25,C22) - 1) / 2	=IF(B22="S",D22,0)
23	188	S	=IF(B23="S",A23-142,A23)	=RANK(C23,$C$2:$C$25,1) +(COUNTIF($C$2:$C$25,C23) - 1) / 2	=IF(B23="S",D23,0)
24	187	S	=IF(B24="S",A24-142,A24)	=RANK(C24,$C$2:$C$25,1) +(COUNTIF($C$2:$C$25,C24) - 1) / 2	=IF(B24="S",D24,0)
25	177	S	=IF(B25="S",A25-142,A25)	=RANK(C25,$C$2:$C$25,1) +(COUNTIF($C$2:$C$25,C25) - 1) / 2	=IF(B25="S",D25,0)
26			Sum =	299	=SUM(E2:E25)

6.4 Interpretation of Test Results

Once the results of the statistical tests are obtained, the specific steps required to achieve site release will depend on the procedures described in the regulatory guide. The following are suggested considerations for the interpretation of the test results with respect to the release limit established for the site.

6.4.1 If the Null Hypothesis Is Not Rejected

Whenever the null hypothesis is not rejected, it is important to complete the analysis by performing a retrospective power analysis for the test. In Scenario A, this will ensure that further remediation is not required simply because the final status survey was not sensitive enough to detect the difference in mean radioactivity concentration between the survey unit and the reference area when that difference is below the LBGR. In Scenario B, this will ensure that a survey unit is not released simply because the final status survey was not sensitive enough to detect the difference in mean radioactivity concentration between the survey unit and the reference area when that difference is above the guideline level. The power analysis may be performed as indicated in Chapter 10, using the actual values of the number of measurements, N, and their observed measurement standard deviation s in place of σ. In some cases, a site-specific simulation of the retrospective power may be warranted when sufficient power cannot be demonstrated by any of the other suggested methods.

If the null hypothesis for the WRS test is not rejected in Scenario B, the Quantile test described in Chapter 7 must also be performed.

6.4.2 If the Null Hypothesis Is Rejected

If the null hypothesis for the Sign test is rejected in Scenario A, it indicates that the residual radioactivity in the survey unit is less than the $DCGL_w$. However, it may still be necessary to document the concentration of residual radioactivity. It is generally best to use the difference in mean radioactivity concentration between the survey unit and the reference area for this purpose.

If the null hypothesis is rejected in Scenario B, it indicates that the residual radioactivity in the survey unit exceeds the LBGR. In this case it is important to determine not only the difference in mean radioactivity concentration between the survey unit and the reference area, δ, but also whether this difference exceeds the release criteria. When the data are normally distributed, the average concentration is generally the best estimator for δ. However, when the data are not normally distributed, other estimators are often better for the same reasons that nonparametric tests are often better than the corresponding parametric tests. These methods are discussed by Lehmann and D'Abrera (1975). When the estimate for δ is below $DCGL_w$, the survey unit may be judged sufficiently remediated, subject to ALARA considerations. Otherwise, further remediation will generally be required.

7 QUANTILE TEST

The Quantile test was specifically developed to detect differences between the survey unit and the reference area that consist of a shift, δ', to higher values in only a fraction, ϵ, $0 < \epsilon < 1$, of the survey unit. It should be noted that, in general, this shift, δ', is not necessarily the same as the shift δ used for the WRS test. The Quantile test is only used in Scenario B. The Quantile test is performed after the WRS test, if the null hypothesis for that test has not been rejected. Using the Quantile test in tandem with the WRS test in Scenario B results in higher power to detect survey units that have not been adequately remediated than either test has by itself.

7.1 Introduction

The specific hypothesis tested by the Quantile test (see Johnson et al., 1987; EPA 230-R-94-004, 1994) is:

Null Hypothesis:
H_0: $\epsilon = 0$ or $\delta' \leq$ LBGR
versus
Alternative Hypothesis:
H_a: $\epsilon > 0$ and $\delta' >$ LBGR

Simply put, the null hypothesis is that there is no residual radioactivity above the LBGR in any part of the survey unit. The Quantile test is better at detecting situations in which only a portion, ϵ, of the survey unit contains excess residual radioactivity. The WRS test is better at detecting situations in which any excess residual radioactivity is uniform across the entire survey unit.

7.2 Applying the Quantile Test

For the Quantile test, the appropriate page in Table A.7 is selected, according to the value of $\alpha_Q = \alpha/2$.[1] Find the nearest values of n, the number of measurements from the survey unit, and m, the number of measurements from the reference area, that are tabulated. There are three numbers associated with each tabulated pair of n and m values, namely r, k, and α_Q.

The m measurements from the reference area and the n adjusted measurements from the survey unit are pooled and ranked in order of increasing size from 1 to N, where $N = m + n$. This is the same as steps (1) – (3) in Section 6.3 for the WRS test, and the same rankings can be used. If k or more of the r largest measurements in the combined ranked data set are from the survey unit, the null hypothesis is rejected For a survey unit that has failed the WRS test, it is not usually necessary to also perform the Quantile test. It is done here as an illustration of the method.

Table 7.1 reproduces the data in Table 6.5, but with two added columns showing the sorted ranks of the adjusted data, and the location associated with each rank, i.e., R for reference area and S for survey unit.

---

[1] Recall that since the Quantile test is performed in tandem with the WRS test, $\alpha_Q = \alpha_w \alpha/2$, so that the that the size of the two tests in tandem is approximately $\alpha = \alpha_Q + \alpha_w$.

Table 7.1 Quantile Test Under Scenario B for Class 2 Interior Drywall Survey Unit
(Measurements from the reference area and the survey unit are denoted by R and S, respectively)

	A	B	C	D	E	F	G
1	Data	Area	Adjusted Data	Ranks	Survey Unit Ranks	Sorted Ranks	Location Associated With Sorted Rank
2	47	R	47	18	—	1	R
3	28	R	28	1	—	2	R
4	36	R	36	6	—	3	R
5	37	R	37	7	—	4.5	R
6	39	R	39	9.5	—	4.5	S
7	45	R	45	13	—	6	R
8	43	R	43	11	—	7	R
9	34	R	34	3	—	8	S
10	32	R	32	2	—	9.5	R
11	35	R	35	4.5	—	9.5	R
12	39	R	39	9.5	—	11	R
13	51	R	51	21	—	13	R
14	209	S	67	24	24	13	S
15	197	S	55	23	23	13	S
16	188	S	46	16	16	16	S
17	191	S	49	19	19	16	S
18	193	S	51	21	21	16	S
19	187	S	45	13	13	18	R
20	188	S	46	16	16	19	S
21	180	S	38	8	8	21	R
22	193	S	51	21	21	21	S
23	188	S	46	16	16	21	S
24	187	S	45	13	13	23	S
25	177	S	35	4.5	4.5	24	S
26			Sum =	300	194.5		

On the second page of Table A.7, the closest entry to $n = m = 12$ is for $n = m = 10$. The values of $r = 7$, $k = 6$ and $\alpha_Q = 0.029$ are found. Thus, the null hypothesis is rejected if 6 of the 7 largest adjusted measurements come from the survey unit. From Table 7.1, we find that only 5 of the 7 largest adjusted measurements come from the survey unit.

The Quantile test as applied above yields only an approximate result. The values of n and m that were used are close to, but not equal to, the actual values. The value of α_Q will generally be different from that listed in the table. It is prudent to check a few other entries that are near the actual sample size. For $n = m = 15$, the values of $r = 5$, $k = 5$, and $\alpha_Q = 0.021$ are found. Thus, the null hypothesis is rejected if all of the 5 largest adjusted measurements come from the survey unit. From Table 7.1, we find that only 4 of the 5 largest adjusted measurements come from the survey unit. For $n = 15$, $m = 10$ the values of $r = 6$, $k = 6$, and $\alpha_Q = 0.028$ are found. From Table 7.1, we find that only 5 of the 6 largest adjusted measurements come from the survey unit. For $n = 10$, $m = 15$, we have $r = 6$, $k = 5$, and $\alpha_Q = 0.023$. Since 5 of the 6 largest adjusted measurements come from the survey unit, the null hypothesis would be rejected in this case. If the results are ambiguous, the methods of the next section can be used to fine tune the test for the sample sizes actually used..

The power of the Quantile test is more difficult to evaluate than that for the WRS test, since it depends on the two parameters δ' and ϵ. Setting specific values of these parameters in order to define a specific alternative for evaluating the power would often be speculative at best. It will be assumed that the power of the quantile test will be adequate for a range of values of δ' and ϵ when the sample size has been determined to assure adequate power for the WRS test.

If there are specific values of δ' and ϵ that are identified as being of concern during the historical site assessment, or prior surveys, power estimates can be obtained from tables in EPA 230-R-94-004 (1994), which list the power for various combinations of ϵ, δ'/σ, α_Q, m, n, r, and k. In those tables, only cases where $n = m$ are given. Since the power generally increases with sample size, upper and lower bounds for the power can be estimated when $m \neq n$ by consulting the entries for sample sizes both equal to the smaller or larger of these numbers.

7.3 Calculation of α_Q for the Quantile Test

The Quantile test, as applied in Section 7.2 gives only an approximate result, since the values used for n and m to find r and k were only approximately equal to the sample sizes actually used. Therefore, the actual value of α_Q will be different from that listed in the table. Fortunately, it is relatively easy to calculate the exact value of α_Q for specific values of n, m, k, and r. The number, k, out of the r largest measurements has a *hypergeometric* probability distribution when the null hypothesis is true. The probability that k or more of the r largest measurements are from the survey unit, when there is actually no residual radioactivity in the survey unit is:

$$\alpha_Q = \sum_{i=k}^{r} \frac{\binom{n}{i}\binom{m}{r-i}}{\binom{n+m}{r}}$$

(7-1)

The symbol $\binom{n}{i} = \dfrac{n!}{(n-i)!\, i!}$ is called a binomial coefficient. The symbol $n!$, called n factorial,

is the product of the first n integers, $n! = n(n-1)(n-2)...(3)(2)(1)$. $0!$ is defined as equal to 1.

These calculations can be performed easily with any spreadsheet program that has the hypergeometric function built in. Otherwise, the following approximation, correct to almost three decimal places (Ling and Pratt, 1984), may be used:

$$1 - \alpha_Q = \Phi(z_{1-\alpha_Q}) \tag{7-2}$$

where

$$z_{1-\alpha_Q} = 2\frac{\sqrt{k(m-r+k)} - \sqrt{(n-k+1)(r-k+1)}}{\sqrt{(n+m-1)}}.$$

$\Phi(z)$ is the cumulative distribution function of a standard normal random variable tabulated in Table A.1. For the actual values of n and m, α_Q can be calculated for different combinations of r and k until a value sufficiently near the DQO is found.

Table 7.2 shows the calculations for the example data used in the previous section. For each of the possible values of r considered, i.e., 5, 6, and 7, the values of α_Q are calculated for each possible value of k, from 0 up to r. The value of α_Q closest to the desired value of $\alpha/2 = 0.025$ occurs for r = 5 and k = 5, where $\alpha_Q = 0.0186$. Using these values, and observing that only 4 of the 5 largest measurements are from the survey unit, the null hypothesis is not rejected. The survey unit passes the Quantile test, but it fails the WRS test (see Section 6.3). This may seem paradoxical, but is actually consistent with the patterns of residual radioactivity each test is designed to detect. This is discussed further in the next section.

Table 7.2 Example Calculation of α_Q for the Quantile Test

	A	B	C	D	E	F	G	H	I	J	K
1	$r=$5				$r=$6				$r=$7		
2	$k=$	Prob	α		$k=$	Prob	α		$k=$	Prob	α
3	0	0.0186	1.0000		0	0.0069	1.0000		0	0.0023	1.0000
4	1	0.1398	0.9814		1	0.0706	0.9931		1	0.0320	0.9977
5	2	0.3416	0.8416		2	0.2427	0.9225		2	0.1510	0.9657
6	3	0.3416	0.5000		3	0.3596	0.6798		3	0.3146	0.8146
7	4	0.1398	0.1584		4	0.2427	0.3202		4	0.3146	0.5000
8	5	0.0186	0.0186		5	0.0706	0.0775		5	0.1510	0.1854
9					6	0.0069	0.0069		6	0.0320	0.0343
10									7	0.0023	0.0023
11		mean k =	2.50			mean k =	3.00			mean k =	3.50
12		std dev =	1.02			std dev =	1.08			std dev=	1.14

The spreadsheet formulas used for the example in Table 7.2 are shown in Table 7.3. In rows 1–12, only the formulas for columns A, B, and C are given, showing the calculations for r = 5, provided the values of n and m are defined in the spreadsheet. The formulas for the other columns are similar. One can get a feeling for the likelihood of the observed value of k by calculating the mean and standard deviation for k when the null hypothesis is true. For $n = m = 12$, and $r = 5$, the expected value of k is 2.5 ± 1.0. Thus, the observed value, $k = 4$, is 1.5 standard deviations above the mean. This is on the high end, but not quite high enough to reject the null hypothesis.

Table 7.3 Spreadsheet Formulas Used in Table 7.2

	A	B	C
1	$r=$	5	
2	$k=$	Prob	α
3	0	=HYPGEOMDIST(A3, *n, r, n+m*)	=1
4	1	=HYPGEOMDIST(A4, *n, r, n+m*)	=1-SUM(B$3:B3)
5	2	=HYPGEOMDIST(A5, *n, r, n+m*)	=1-SUM(B$3:B4)
6	3	=HYPGEOMDIST(A6, *n, r, n+m*)	=1-SUM(B$3:B5)
7	4	=HYPGEOMDIST(A7, *n, r, n+m*)	=1-SUM(B$3:B6)
8	5·	=HYPGEOMDIST(A8, *n, r, n+m*)	=1-SUM(B$3:7)
9			
10			
11		mean $k=$	$= (n^{*}r) / (m+n)$
12		std dev $=$	$= sqrt(\ m^{*}n^{*}r^{*}(m+n\text{-}1) / (m+n)^2\ )$

7.4 Modified Example for the Quantile Test

It was noted in the previous section that for the example data of Table 7.1 under Scenario B, the survey unit would fail the WRS test, yet pass the Quantile test. This occurred because the residual radioactivity is more or less uniformly distributed across the survey unit. If the example data is modified slightly, the result is different. Table 7.4 shows what would happen if 15 units of residual radioactivity are subtracted from the first six survey unit measurements and added to the last six survey unit measurements. The total residual radioactivity measured is unchanged. The analysis shows, however, that the sum of the adjusted survey unit measurement ranks is now 178. This is below the critical value of 184 from Section 6.3, and so the survey unit would pass the WRS test. Now, however, all of the highest ranked five measurements are from the survey unit, and so the survey unit would fail the Quantile test. This is because the spatial distribution of residual radioactivity is not uniform over the survey unit, but is concentrated at higher values in half of the survey unit. The purpose of using the two tests in tandem under Scenario B is to discover survey units with residual radioactivity in excess of the LBGR. There would be no value in using two tests if they always gave the same result.

Table 7.4 Quantile Test Under Scenario B: Modified Example Survey Unit
(Measurements from the reference area and the survey unit are denoted by R and S, respectively)

	A	B	C	D	E	F	G
1	Data	Area	Adjusted Data	Ranks	Survey Unit Ranks	Sorted Ranks	Location Associated With Sorted Rank
2	47	R	47	16	—	1	R
3	28	R	28	1	—	2	S
4	36	R	36	8.5	—	3	S
5	37	R	37	10	—	4	R
6	39	R	39	11.5	—	5.5	R
7	45	R	45	15	—	5.5	S
8	43	R	43	14	—	7	R
9	34	R	34	5.5	—	8.5	R
10	32	R	32	4	—	8.5	S
11	35	R	35	7	—	10	R
12	39	R	39	11.5	—	11.5	R
13	51	R	51	18	—	11.5	R
14	194	S	52	19	19	13	S
15	182	S	40	13	13	14	R
16	173	S	31	3	3	15	R
17	176	S	34	5.5	5.5	16	R
18	178	S	36	8.5	8.5	17	S
19	172	S	30	2	2	18	R
20	203	S	61	22.5	22.5	19	S
21	195	S	53	20	20	20	S
22	208	S	66	24	24	21	S
23	203	S	61	22.5	22.5	22.5	S
24	202	S	60	21	21	22.5	S
25	192	S	50	17	17	24	S
26			Sum =	300	178		

8 ELEVATED MEASUREMENT COMPARISON

As discussed in Section 2.6, an *Elevated Measurement Comparison* is performed by comparing each measurement from the survey unit to the $DCGL_{EMC}$. If the survey unit is being compared to a reference area, the net survey unit measurement is first obtained by subtracting the mean of the reference area measurements. A net survey unit measurement that equals or exceeds the $DCGL_{EMC}$ is an indication that a survey unit may contain residual radioactivity in excess of the release criterion.

In addition to direct measurements or samples at discrete locations, parts of each survey unit will also be scanned. For the quantitative measurements obtained at discrete locations, performing the EMC is a straightforward comparison of two numerical values. Some sophisticated scanning instrumentation is also capable providing quantitative results with a quality approaching those from direct measurements or samples. Other scanning measurements, however, may be more qualitative. In that case, *action levels* should be established for the scanning procedure so that areas with concentrations that may exceed the $DCGL_{EMC}$ are marked for a quantitative measurement.

8.1 Introduction

The Elevated Measurement Comparison (EMC) against measurements taken on a systematic grid are discussed in Section 8.1. The use of the EMC during scans is discussed in Section 8.2. Area factors are discussed in Section 8.3, and an example is given in Section 8.4.

The statistical tests may not fail a survey unit when there are only a very few high measurements. The EMC is used so that unusually large measurements will receive proper attention regardless of the outcome of those tests—and any area that may have the potential for significant dose contributions will be identified. The EMC is intended to flag potential failures in the remediation process, and cannot be used to determine whether or not a site meets the release criterion until further investigation is done.

The derived concentration guideline level for the EMC is: $DCGL_{EMC} = (F_{grid})(DCGL_W)$, where F_{grid} is the area factor for the area of the systematic grid area used (see Section 3.5.4). Note that $DCGL_{EMC}$ is an *a priori* limit, established both by the $DCGL_W$ and by the survey design (i.e., grid spacing and scanning MDC). The true extent of an area of elevated activity can only be determined after performing the survey and then taking additional measurements if an elevated measurement is found. Upon the completion of further investigation, the *a posteriori* limit, $DCGL_{EA} = (F_{actual})(DCGL_W)$, can be established using the value of the area factor, F, appropriate for the *actual measured area of elevated concentration*. The area that is considered elevated is that bounded by concentration measurements at or below the $DCGL_W$.

If residual radioactivity is found in an isolated area of elevated activity—in addition to residual radioactivity distributed relatively uniformly across the survey unit—the unity rule can be used to ensure that the total dose is within the release criterion:

$$\frac{\delta}{DCGL_W} + \frac{(\text{average concentration in elevated area} - \delta)}{(\text{area factor for elevated area})(DCGL_W)} < 1$$

If there is more than one elevated area, a separate term should be included for each. As an alternative to the unity rule, the dose or risk due to the actual residual radioactivity distribution can be calculated if there is an appropriate exposure pathway model available.

The preceding discussion primarily concerns Class 1 survey units. Measurements exceeding $DCGL_W$ in Class 2 or Class 3 areas may indicate survey unit mis-classification. Scanning coverage requirements for Class 2 and Class 3 survey units are less stringent than for Class 1 survey units.

If the investigation levels of Section 2.6 are exceeded, an investigation should (1) assure that the area of elevated activity discovered meets the release criterion and (2) provide reasonable assurance that other undiscovered areas of elevated activity do not exist. If further investigation determines that the survey unit was misclassified with regard to contamination potential, a resurvey using the method appropriate for the new survey unit classification may be appropriate.

8.2 Comparison Against Individual Measurements

The $DCGL_{EMC}$ is calculated on the basis of the grid area, since this is about the same as the largest circular area that has some chance of being missed when sampling on the grid. Figure 8.1 shows both a square and a triangular sampling grid. On the square grid, with grid area L^2, the small circular area with diameter L has an area of $\pi(L/2)^2 = 0.785L^2$. A circle with area L^2 would have a radius of $0.564L$. From Figure 3.8, a circle of that radius has only about a 10% chance of being missed. The triangular grid is a little more efficient. The grid area is the rhombus formed from two of the triangles, and has area of $0.866L^2$. A circle with that area has a radius of $0.525L$, and, from Figure 3.8, has less than a 5% chance of being missed. The significance of this is that when no measurement exceeds the $DCGL_{EMC}$ it is unlikely that there are areas remaining that could cause the release criterion to be exceeded. The survey is planned in anticipation of a negative result, and provides a quantitative measure of risk when no elevated measurements are found, as well as an objective, dose-based definition of what is considered elevated.

When a measurement is found to exceed the $DCGL_{EMC}$ there is more work to be done before the question of compliance can be answered. An individual elevated measurement on a systematic grid could conceivably represent an area three to four times as large as the systematic grid area used to define the $DCGL_{EMC}$. This is the area bounded by the nearest neighbors of the elevated measurement location, as shown by the large circles in Figure 8.1. However, the elevated area may also be smaller than the grid area. Since the allowable concentration generally increases as the elevated area decreases in size, further investigation is necessary to determine both the actual area and average concentration. The boundary of the elevated area is defined by concentration measurements at or below the $DCGL_W$. Once the actual elevated area is found, the corresponding area factor, F_{actual} is calculated in order to determine the release criterion for the elevated area: $DCGL_{EA} = (F_{actual})(DCGL_W)$.

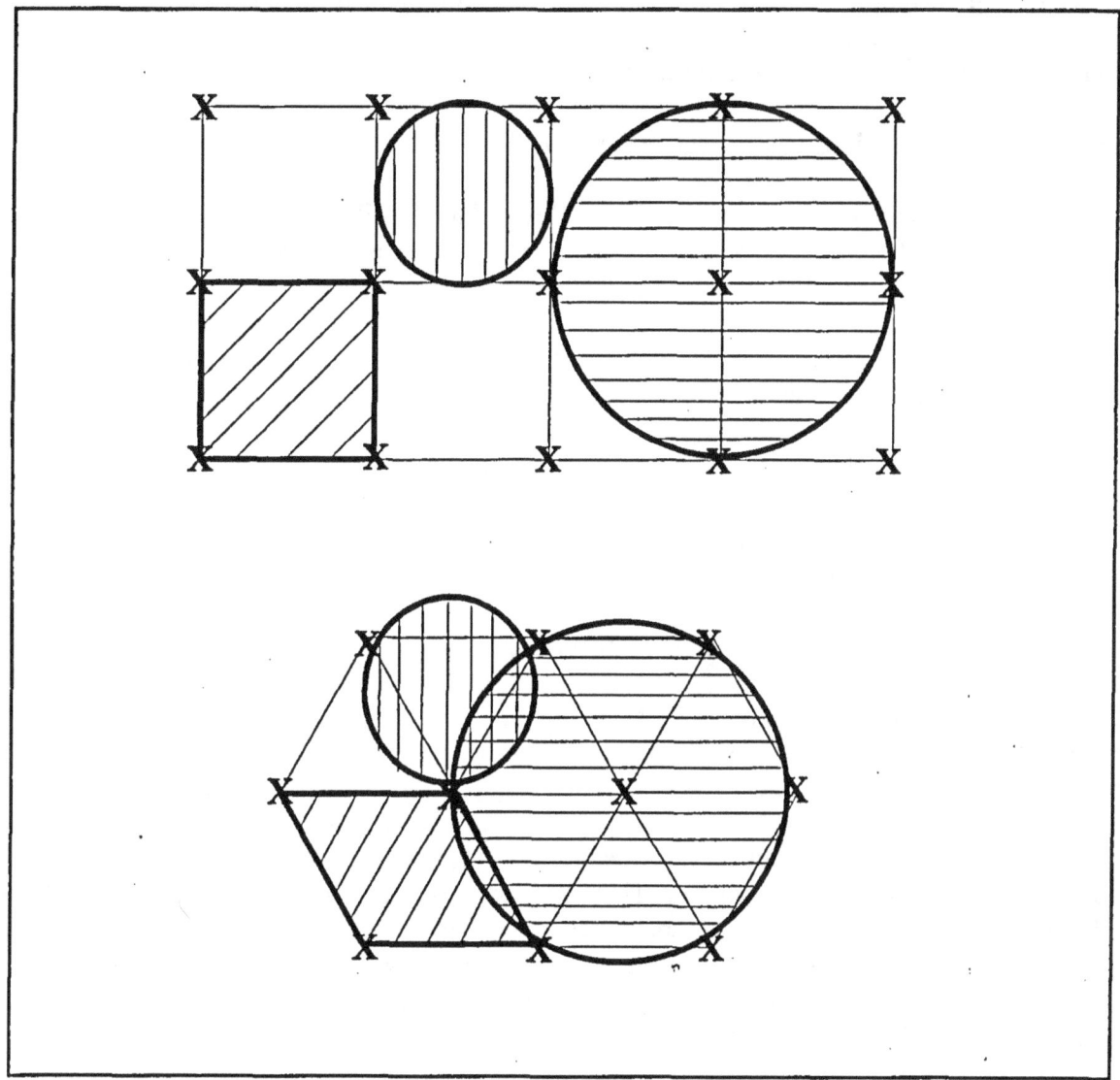

Figure 8.1 Square (top) and Triangular (bottom) Sampling Grids and Grid Areas
(Circular elevated areas of radius $L/2$ and L are shown for comparison)

The problem remaining is to determine whether or not the average concentration in the elevated area meets the $DCGL_{EA}$. This is essentially the same problem as the original one of determining whether or not the average concentration in the survey unit meets the $DCGL_W$. This is not to suggest that it is necessary to define the elevated area as a separate survey unit and conduct a new survey to determine its compliance with the $DCGL_{EA}$. For cases in which the decision is too close to call, it may be useful to keep this analogy in mind when planning a resolution to the problem. It will also be useful in planning the investigation of the elevated area.

There will be many types and sizes of elevated areas. In many cases, it may be obvious whether or not the elevated area exceeds the release criterion based on the measurements taken during the investigation, without performing an additional survey or performing additional statistical tests. Obviously, if the elevated area mean exceeds the $DCGL_{EA}$, the survey unit fails. If the elevated area mean is less than the $DCGL_{EA}$ by more than the standard error of the mean, ALARA considerations will usually determine whether or not further remediation is necessary. As with any measurements, the DQO process should be used in planning the investigation of the elevated area, and what decisions will be made based on the results.

Some other considerations that may arise are:

(1) The variability of concentrations in the elevated area is likely to exceed that of any background variations, so additional reference area measurements will not usually be needed. If the survey unit is being compared to a reference area, the boundary of the elevated area should be determined by measurements at or below the $DCGL_W$ added to the mean reference area measurement. The elevated area mean minus the mean reference area concentration should not exceed the $DCGL_{EA}$.

(2) There may be elevated areas within the elevated area: There may exist a smaller area within the elevated area that has concentrations high enough to exceed the release criterion when considered separately, even though the average concentration over the entire elevated area is within the $DCGL_{EA}$.

8.3 Comparison Against Scanning Measurements

The measurement results obtained during scanning are inherently more qualitative in nature than those obtained on the systematic grid. In Class 1 survey units, much of the survey design depends on the ability to detect areas exceeding the $DCGL_{EMC}$ during scanning. This is the essence of the requirement that the scanning MDC be below the $DCGL_{EMC}$. In practice, this means that an operating procedure for flagging suspect locations during scanning be devised to ensure that potential elevated areas be investigated. Then, as is the case with measurements on the systematic grid that exceed the $DCGL_{EMC}$, the suspect area must be investigated to determine the area and average concentration of the elevated area. In many cases, it would be prudent to set the criteria for flagging elevated areas conservatively. If this is done, and subsequent quantitative measurements indicate that the $DCGL_{EMC}$ is not actually being exceeded, nothing further would generally be necessary unless for ALARA considerations. If measurements above the stated scanning MDC are found by sampling or by direct measurement at locations that were not flagged by the scanning survey, this may indicate that the scanning method did not meet the DQOs.

Scanning requirements for Class 2 and Class 3 survey units are less stringent both with regard to the coverage and sensitivity. This is possible because of the screening process necessary to show that these areas are not highly contaminated (Sections 2.2.2 and 2.2.3). For this reason, the investigation levels are lower than in Class 1 areas (Section 2.5.7).

8.4 Area Factors

Area factors have been discussed in Sections 2.2.1 and 3.8.2. These area factors should be calculated using dose pathway models and assumptions that are consistent with those used to calculate the $DCGL_W$. In this section, examples of area factors for both indoor and outdoor survey units are given.

The example outdoor area factors listed in Table 8.1 were calculated using RESRAD for Windows 5.70 (ANL/EAD/LD-2). For each radionuclide, all dose pathways were calculated assuming an initial concentration of 1 pCi/g. The default area of contamination in RESRAD 5.7 is 10000 m^2, so for this size area, the area factor for all radionuclides is equal to one. Area factors for other size areas were computed by taking the ratio of the dose per unit concentration calculated by RESRAD for the default 10000 m^2 to that calculated for 1, 3, 10, 30, 100, 300, 1000, and 3000 m^2. The other RESRAD default values were not changed except to adjust the length parallel to aquifer to be consistent with area of contamination..

The area factors for selected radionuclides are plotted in Figure 8.2. There it can be seen that radionuclides generally fall into three groups. Those that deliver dose primarily through internal pathways, those that deliver dose primarily through the external pathway, and a few for which both are important. Generally, the radionuclides that deliver dose via internal pathways (e.g., ^{14}C, ^{90}Sr) have the highest area factors. These area factors scale with the area in a manner suggesting that it is the total inventory of the radionuclides that is most important. The area factors for radionuclides that deliver dose primarily through external gamma have lower area factors, reflecting the fact that these radionuclides can deliver dose at a distance. In a mixture, it will generally be these radionuclides that will have the limiting area factors. Fortunately, these are also the radionuclides most easily detected using scanning techniques.

Notice that Figure 8.2 is plotted on a log-log scale. Linear interpolation on this figure corresponds to logarithmic interpolation in Table 8.1 for areas between those listed. For example, if the area factor for ^{241}Am is needed for 25 m^2, the table lists 96.3 for 10 m^2 and 44.2 for 30 m^2. To interpolate, take the base 10 logarithms of these numbers:

$$\log_{10}(10) = 1$$
$$\log_{10}(30) = 1.477$$
$$\log_{10}(25) = 1.398$$
$$\log_{10}(13.4) = 1.127$$
$$\log_{10}(4.99) = 0.698.$$

The interpolation is done using these values:

$$\log_{10}(A_{25}) = \log_{10}(13.4)$$
$$+ [\log_{10}(25) - \log_{10}(10)] \{ [\log_{10}(4.99) - (\log_{10}(13.4)] / [\log_{10}(30) - \log_{10}(10)]\}$$

$$= 1.127 + [1.398 - 1)] \{ [0.698 - 1.127] / [1.477 - 1]\}$$

$$= 1.127 + [0.398] \{ [-0.429] / [0.477]\}$$

$$= 0.769$$

Therefore, $A_{25} = 10^{(0.769)} = 5.88$.

Example indoor area factors listed in Table 8.2 were calculated using RESRAD BUILD for Windows 2.11 (ANL/EAD/LD-3, 1994). For each radionuclide, all dose pathways were calculated assuming an initial concentration of 1 pCi/m^2. The default area of contamination in

RESRAD BUILD is 36 m². The other areas compared to this value were 1, 4, 9, 16, or 25 m². No other changes to the RESRAD BUILD default values were made. Dose was computed for one receptor, who spent 100% of time in the contaminated room. The area factors were then computed by taking the ratio of the dose per unit concentration calculated by RESRAD BUILD for the default 36 m² to that calculated for the other areas listed. Thus, if the guideline limit concentration for residual radioactivity distributed over 36 m² is multiplied by this value, the resulting concentration distributed over the specified smaller area delivers the same average dose. There are obviously many other exposure scenarios which may result in different area factors.

Table 8.1 Example Outdoor Area Factors

Nuclide	10000m²	3000 m²	1000 m²	300 m²	100 m²	30 m²	10 m²	3 m²	1 m²
Am-241	1.00	1.01	1.01	1.20	1.86	4.99	13.4	40.2	109
C-14	1.00	2.09	3.06	4.84	8.40	23.6	65.7	207	609
Cd-109	1.00	1.03	1.04	1.11	1.42	3.05	7.61	22.1	63.0
Ce-144	1.00	1.03	1.04	1.11	1.21	1.49	2.05	4.24	9.30
Co-57	1.00	1.03	1.05	1.11	1.19	1.46	1.99	4.06	8.69
Co-60	1.00	1.04	1.06	1.13	1.23	1.52	2.12	4.39	9.81
Cs-134	1.00	1.07	1.10	1.19	1.30	1.61	2.22	4.57	10.1
Cs-137	1.00	1.10	1.14	1.28	1.41	1.75	2.41	4.98	11.0
Eu-152	1.00	1.03	1.05	•1.10	1.19	1.47	2.03	4.20	9.28
Fe-55	1.00	2.12	3.12	9.97	27.1	71.6	149	284	484
H-3	1.00	1.06	1.08	1.38	2.18	5.97	16.4	51.3	150
I-129	1.00	1.19	1.34	1.90	3.14	8.92	25.0	79.1	233
Mn-54	1.00	1.03	1.05	1.12	1.22	1.50	2.08	4.30	9.52
Na-22	1.00	1.05	1.08	1.13	1.22	1.51	2.07	4.28	9.44
Nb-94	1.00	1.01	1.05	1.18	1.27	1.56	2.15	4.42	9.77
Ni-63	1.00	1.46	1.68	5.59	16.6	54.2	155	464	1180
Pu-238	1.00	1.02	1.04	1.82	2.50	3.26	4.24	6.01	8.88
Pu-239	1.00	1.02	1.03	1.83	2.51	3.28	4.26	6.07	8.94
Ru-106	1.00	1.01	1.02	1.07	1.36	2.65	7.23	14.9	32.7
Sb-125	1.00	1.03	1.05	1.10	1.18	1.45	1.99	4.10	8.94
Sr-90	1.00	1.17	1.23	4.04	11.9	37.1	98.7	285	729
Tc-99	1.00	1.02	1.07	1.54	2.55	7.16	20.0	62.8	185
Th-232	1.00	1.03	1.05	1.47	1.75	2.24	3.12	6.08	12.3
U-235	1.00	1.01	1.19	2.18	3.84	10.3	15.9	30.2	58.8
U-238	1.00	1.01	1.04	1.43	2.27	5.73	11.1	18.3	30.5
Zn-65	1.00	1.31	1.45	1.81	2.07	2.62	3.64	7.62	17.0

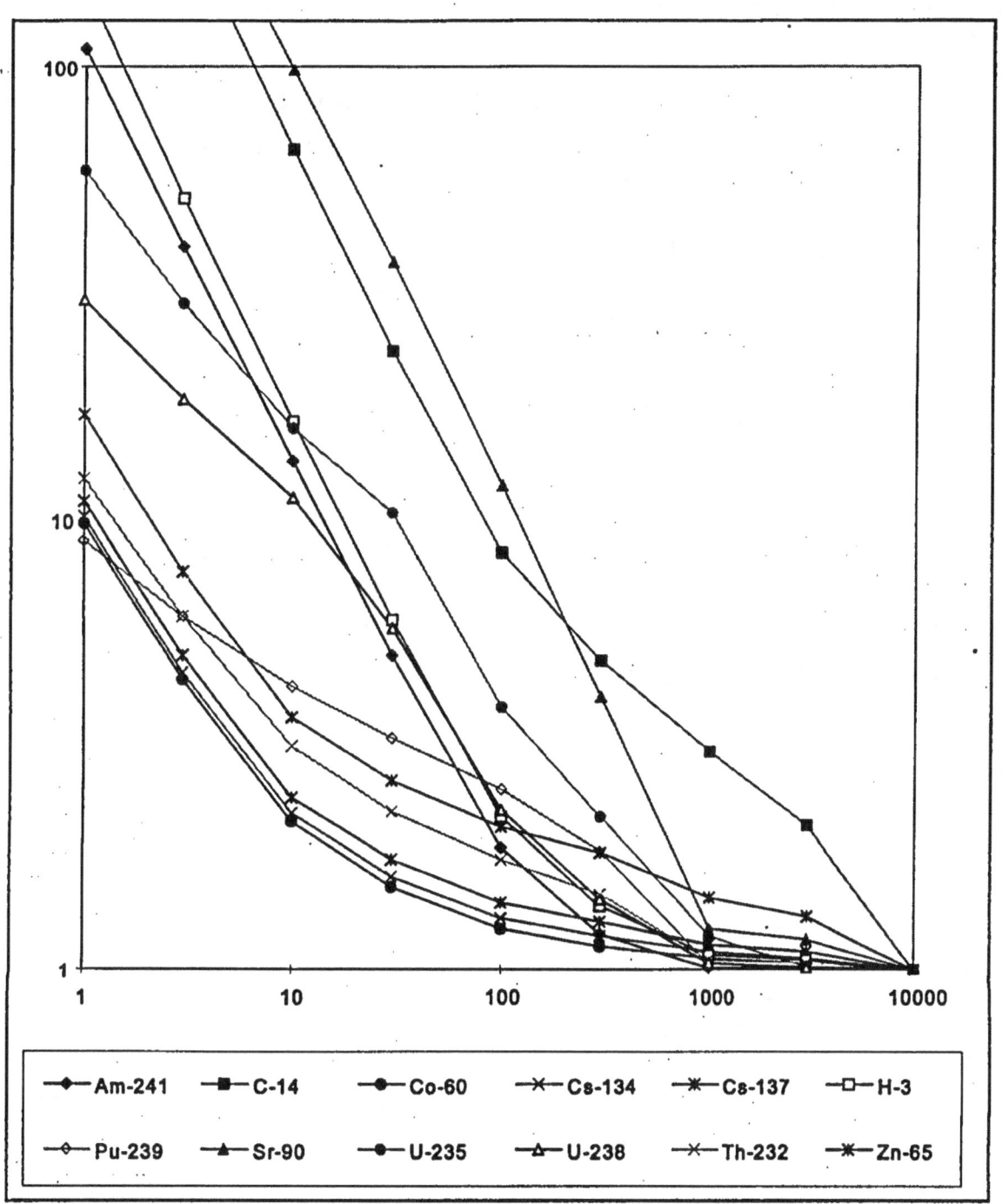

Figure 8.2 Example Outdoor Area Factors

The indoor area factors for selected radionuclides are plotted in Figure 8.3. There is not as much variation between radionuclides as there is with the outdoor area factors. All of the area factors scale nearly with the size of the contaminated area. As with the outdoor area factors, the radionuclides that deliver dose primarily through internal pathways have higher area factors than those that deliver dose primarily through the external pathway. The area factors for radionuclides that deliver dose primarily through internal pathways scale with the area in a manner suggesting that it is the total inventory of the removable fraction of these radionuclides that is most important.

Table 8.2 Example Indoor Area Factors

Nuclide	36 m²	25 m²	16 m²	9 m²	4 m²	1 m²
Co-60	1.0	1.2	1.6	2.5	5.5	22.7
Cs-137	1.0	1.2	1.7	2.6	5.7	23.5
H-3	1.0	1.4	2.3	4.0	9.0	36.0
Ni-63	1.0	1.4	2.3	4.0	9.0	36.0
Pu-239	1.0	1.4	2.3	4.0	9.0	36.0
Ra-226	1.0	1.4	2.1	3.6	8.0	32.2
Sr-90	1.0	1.4	2.2	4.0	8.9	35.7
Th-232	1.0	1.4	2.3	4.0	9.0	36.0
U-235	1.0	1.4	2.2	4.0	9.0	35.8
Zn-65	1.0	1.2	1.6	2.5	5.4	22.3

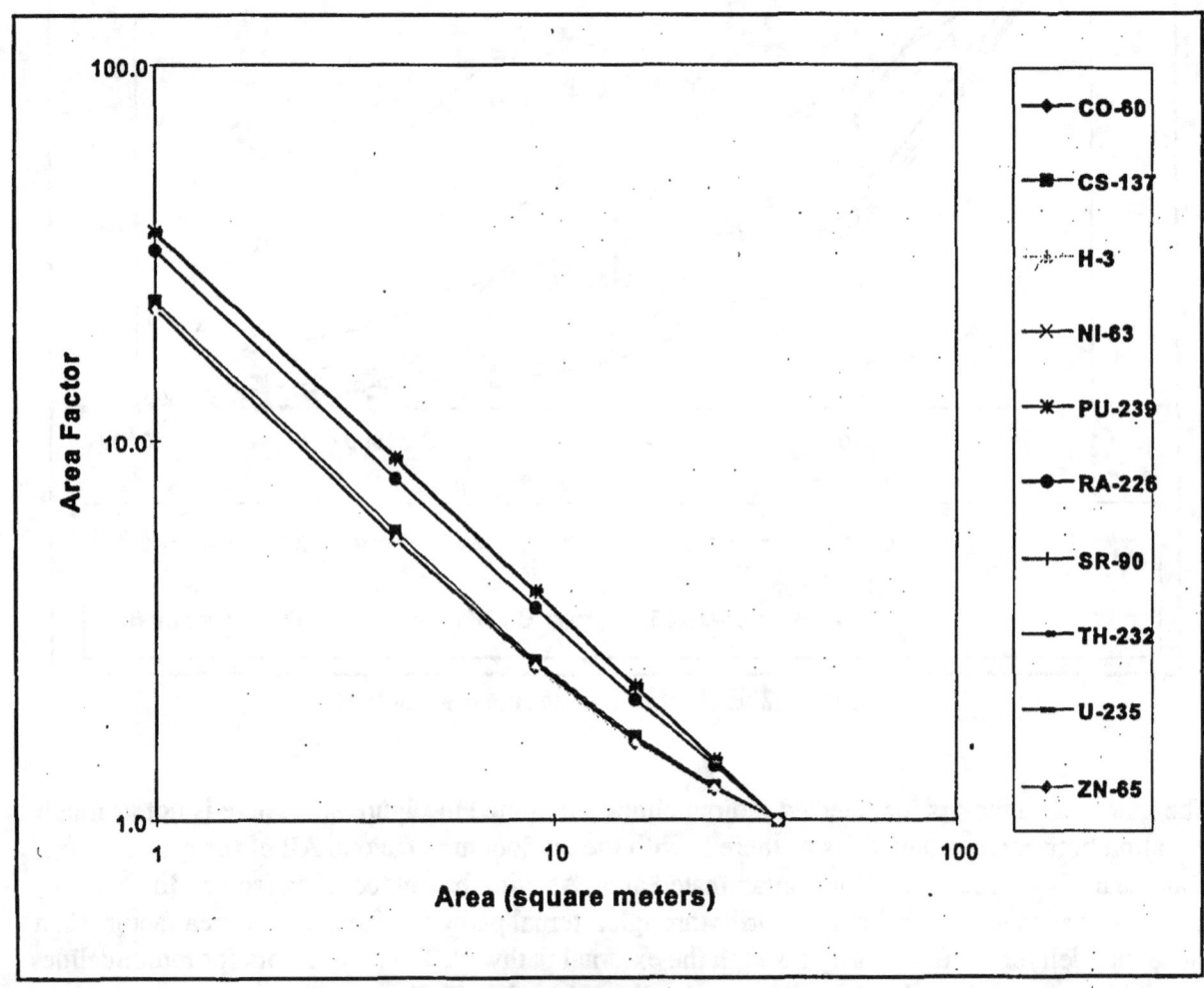

Figure 8.3 Example Indoor Area Factors

The area factors for radionuclides that deliver dose primarily through external gamma have lower area factors, reflecting the fact that this dose can be delivered at a distance. Thus, in a mixture, it will generally be these radionuclides that will usually have the limiting area factors. However, the effect is not as large indoors as outdoors.

8.5 Example

A concrete room, 5 meters by 6 meters, had been contaminated with ^{137}Cs and ^{60}Co, and subsequently remediated. The floor and the bottom 2 meters of the walls were to be surveyed as a Class 1 survey unit. Measurements were to be made for 100s at each grid point with a 16 cm^2 GM counter with a 10.1% efficiency. The DCGL$_W$ for both nuclides were within about 10%, so the lower was taken to conservatively apply to both. This DCGL$_W$, about 1100 dpm per 100 cm^2, translated into 30 counts per 100s with this detector. During the DQO process it was determined that Scenario A would be used with $\alpha = 0.05$, $\beta = 0.025$, and $\Delta = 10$ counts. The average background readings for this type of building on site had been about 60 ± 10 counts, so the estimated $\sigma = 10$. From Table 3.3, it was found that 39 measurements each were required in the reference area and the survey unit. This was rounded up to 40. The survey unit area is 61 m^2, so the spacing, L, on a triangular grid is $L = [61/ (0.866N)]^{1/2} = [61/ (34.6)]^{1/2} = 1.3$ m, using $N = 40$. The grid area is $0.866 L^2 = 0.866 (1.3)^2 = 1.5$ m^2. Interpolating into Table 8.2 gives an area factor for 1.5 m^2 of 15. This results in a $DCGL_{EMC} = 15 (DCGL_W) = 16500$ dpm per 100 cm^2, or 450 GM counts per 100s. This level is easily seen while scanning, so no additional grid measurements will be needed in order to find elevated areas.

When the random start triangular grid was laid out in the survey unit, 50 measurement locations were identified. When there are more locations identified than are required, they are all sampled and reported. In the reference area, the grid lay out was terminated when 40 locations were found. The data are shown in Table 8.3.

The mean and standard deviation of the reference area measurements was 58 ± 10. For the survey unit, these were 88 ± 92. The difference of the means, $88 - 58$, is just at the DCGL$_W$ of 30. *However the median in the reference area is 59 while that in the survey unit is only 58.* This is an indication that the survey unit data is fairly symmetric, but that the survey unit mean is being driven up by a few very high measurements. This can be seen even more clearly in the combined ranked data plot of Figure 8.4. For this plot the reference area measurements, adjusted by adding the DCGL$_W$ to each, are combined with the survey unit measurements. The measurements are then plotted against their rank in the combined data set, using different symbols for the reference area points and the survey unit points. This is an easier diagnostic plot to use than the Quantile-Quantile plot when the reference area and survey unit have different numbers of data points. From this plot it can be seen that the majority of the survey unit measurements fall below the adjusted reference area measurements, but that there are eight survey unit measurements that are much higher. One of those measurements exceeds the DCGL$_{EMO}$ which is equivalent to 450 GM counts per 100s. Thus, further investigation of this survey unit will be required before it could be released, regardless of the outcome of the WRS test.

The sum of the adjusted reference area ranks, shown in Table 8.3, is 2432. This is greater than the critical value of 2023 given by the equation following Table A.3 for $n = 50$, $m = 40$, and $\alpha = 0.05$. Thus, the hypothesis that the survey unit as a whole *uniformly* exceeds the DCGL$_W$ is rejected. Whether or not the survey unit may be released is now dependent on the results of the investigation of the elevated measurements that were found.

A posting plot of the survey unit data is shown in Figure 8.5. In terms of GM counts, the elevated area is defined by the average reference area measurement plus the $DCGL_W$ which is $58 + 30 = 88$. The shaded area in Figure 8.5 encloses the measurements exceeding 88 GM counts per 100s. This area, which also encloses all of the measurements exceeding the $DCGL_{EMO}$ covers almost 16 m^2. From Table 8.2, the area factor for ^{137}Cs is 1.7 and that for ^{60}Co is 1.6. For a mixture of the two radionuclides, the smaller value is used. Thus, in this case the $DCGL_{EA} = 1.6$ ($DCGL_W$) = 1760 dpm per 100 cm^2, or 48 GM counts per 100s. The average of the ten measurements in the shaded area is 216.8 GM counts per 100s, which is $216.8 - 58 = 158.8$ GM counts per 100s above the reference area average. Thus, the survey unit does not meet the release criterion, and may not be released without further remediation.

This example illustrates how the nonparametric statistical tests used in combination with the elevated measurement comparison work to assure that the release criterion is met.

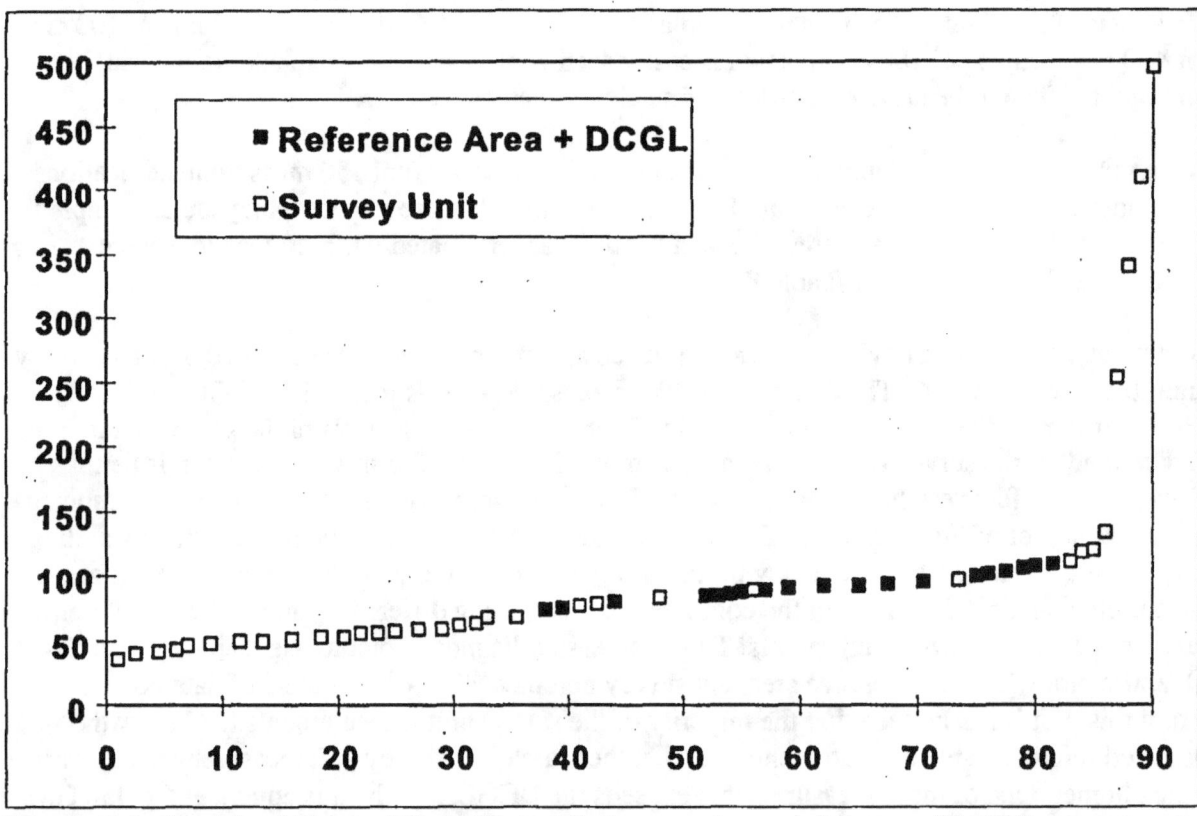

Figure 8.4 Combined Ranked Data Plot of Reference Area and Survey Unit Measurements

Table 8.3 Data for Indoor Survey Unit and Reference Area
(GM counts per 100s)

Ref Data	Survey Data	Adjusted Ref Data	Ref Ranks	Survey Ranks
55	56	88	55	23.5
50	46	83	48	7
63	55	96	70.5	22
51	58	84	52	27
43	68	76	39.5	35.5
50	51	83	48	16
50	83	83	48	48
50	89	83	48	56
50	47	83	48	9
53	134	86	54	86
35	410	68	35.5	89
50	78	83	48	42.5
43	52	76	39.5	18.5
73	40	106	79	2.5
63	111	96	70.5	83
75	340	108	80	88
60	50	93	65	13.5
58	43	91	59	6
70	62	103	77.5	30.5
61	495	94	67.5	90
68	63	101	76	32
57	47	90	57	9
63	52	96	70.5	18.5
60	254	93	65	87
58	78	91	59	42.5
59	67	92	62	33
47	97	80	44	73.5
58	59	91	59	29
76	49	109	81.5	11.5
61	118	94	67.5	84
70	120	103	77.5	85
67	77	100	75	41
60	42	93	65	4.5
76	53	109	81.5	20.5
52	68	85	53	35.5
59	68	92	62	35.5
41	36	74	38	1
63	62	96	70.5	30.5
64	56	97	73.5	23.5
59	58	92	62	27
	47			9
	40			2.5
	50			13.5
	57			25
	58			27
	49			11.5
	53			20.5
	51			16
	42			4.5
	51			16
		Sum =	2432	1663

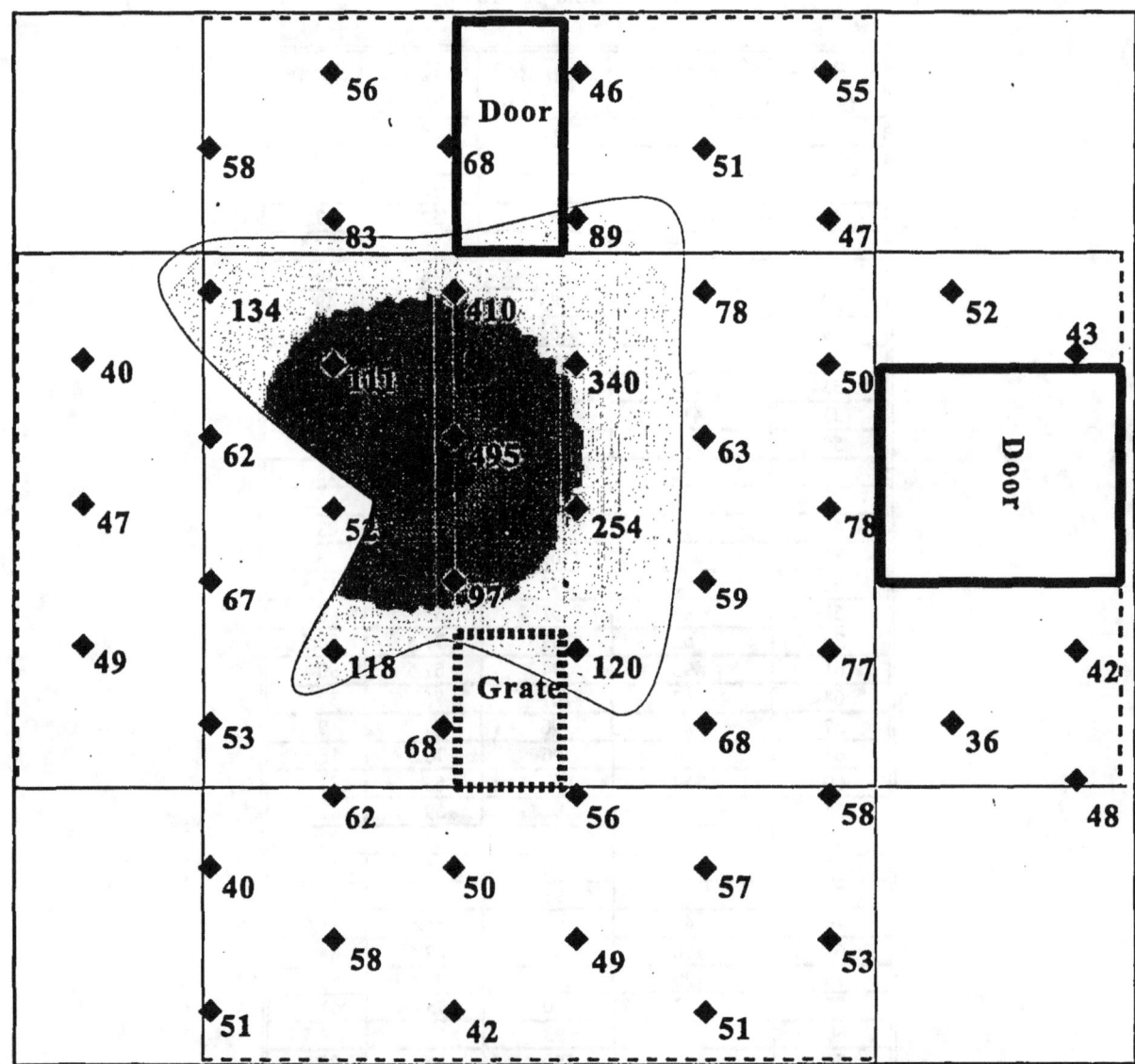

Figure 8.5 Posting Plot of Indoor Concrete Survey Unit

9 SAMPLE SIZE

9.1 Sample Size and Decision Errors

To determine the number of samples to collect, acceptable values of the Type I error rate (α) and Type II error rate (β) must be specified as part of a statistical test. The process for doing this was discussed in Section 3.7. If there are many survey units and each unit requires a separate decision, even if H_0 is true approximately $100\alpha\%$ of the times the test is conducted, the null hypothesis will be incorrectly rejected. If a smaller value of α is used, the number of times this can be expected to happen decreases proportionately. On the other hand, larger values of α will reduce the number of samples initially required from each survey unit.

The power $(1-\beta)$ is the ability of a statistical test to detect when the null hypothesis is indeed false and should be rejected. A test should have high power, i.e., small β, but smaller specified values of β require a larger number of measurements.

The number of samples depends not only on α and β, but also on the width of the gray region relative to the measurement variability, Δ/σ. This parameter essentially describes the resolution of the decision problem. When the resolution is high, only a few measurements are needed. When the resolution is low, many more measurements may be required, even though the specified α and β are unchanged.

The DQO steps described in Sections 3.7 and 3.8 are used to balance the cost of sampling against the risks involved with any potential decision errors. It is important to realize, however, that the value of α is fixed at the desired value when the critical value for the test statistic is determined from the provided tables and used in the test. If the sample size is larger (or smaller) than planned, the effect will be an increase (or decrease) in the power, $1-\beta$. Similarly, if the measurement standard deviation is larger (or smaller) than anticipated, the effect will be a decrease (or increase) in the power.

The consequences of increasing or decreasing the power depends on whether Scenario A or Scenario B is being used. In Scenario A, where the null hypothesis is that the survey unit does not meet the release criterion, high power means that a survey unit that meets the release criterion has a high probability of passing the test. In Scenario B, where the null hypothesis is that the survey unit meets the release criterion, high power means that a survey unit that does not meet the release criterion has a high probability of failing the test.

In most cases, the sample sizes required can be determined using Tables 3.2 and 3.3. In the following sections, the assumptions made and the calculations performed in creating these tables are described. Methods for modifying the calculations under alternative assumptions are also given. It must be emphasized that relatively little effort is required to perform the suggested sample size determinations compared to the time and expense involved in collecting and analyzing samples. This is a key advantage to using the DQO process to determine sample sizes.

9.2 Sample Size Calculation for the Sign Test Under Scenario A

For the Sign test, the number of samples, N, required from the survey unit can be approximated from a formula given by Noether (1987):

$$N = \frac{(Z_{1-\alpha} + Z_{1-\beta})^2}{4(p-0.5)^2} \tag{9-1}$$

where:
α = specified Type I error rate
β = specified Type II error rate
$Z_{1-\alpha}$ = 100(1- α) percentile of the normal distribution
$Z_{1-\beta}$ = 100(1- β) percentile of the normal distribution
p = estimated probability that a random measurement from the survey unit will be less than the DCGL$_w$ when the survey unit median is actually at the LBGR. $p \neq 0.5$

Commonly used values for α and β, and the corresponding values of $Z_{1-\alpha}$ (or $Z_{1-\beta}$) may be found from Table 9.1. Other values can be obtained using any table of the cumulative standard normal distribution function, such as that in Appendix A.

Table 9.1 Some Values of $Z_{1-\alpha}$ and $Z_{1-\beta}$ Used To Calculate the Sample Sizes

α (or β)	$Z_{1-\alpha}$ (or $Z_{1-\beta}$)
0.01	2.3268
0.025	1.9604
0.05	1.6452
0.10	1.2817
0.20	0.8415

The numerator of Equation 9.1, $(Z_{1-\alpha} + Z_{1-\beta})^2$, depends on α and β, but not on Δ or σ. In addition, it only depends on the pair of values used for α and β, and is the same if these values are reversed. This can be seen in the symmetry of Table 9.2, where the value of $(Z_{1-\alpha} + Z_{1-\beta})^2$ is given for each pair of the values of α and β listed in Table 9.1. As will be seen, these are also the minimum samples sizes for each pair of α and β values.

The denominator of Equation 9.1 depends on the parameter p, but not on α or β. The definition. of the parameter p states that it is the estimated probability that a random measurement from the survey unit will be less than the DCGL$_w$ when the survey unit median is actually at the LBGR.

This is illustrated in Figure 9.1. The value of $1-p$ expresses the likelihood that measurements exceeding the DCGL$_w$ will be observed, even if half of the concentration distribution is below the LBGR. This likelihood is higher when the measurement standard deviation is large compared to the width of the gray region. Some assumptions about the data distribution have to be made in estimating p. If it were possible to specify p exactly, there would be no need to do the survey.

Table 9.2 Some Values of $(Z_{1-\alpha} + Z_{1-\beta})^2$ Used To Calculate Sample Sizes

β	α				
	0.01	0.025	0.05	0.1	0.2
0.01	22	19	16	14	11
0.025	19	16	13	11	8
0.05	16	13	11	9	7
0.1	14	11	9	7	5
0.2	11	8	7	5	3

The relative width of the gray region, Δ/σ, is especially useful for estimating the parameter p. If the data are even approximately normally distributed, then p can be estimated from

$$
\begin{aligned}
p &= \frac{1}{\sqrt{2\pi}\,\sigma} \int_{-\infty}^{DCGL_W} e^{-(x-LBGR)^2/2\sigma^2} dx \\
&= \frac{1}{\sqrt{2\pi}\,\sigma} \int_{-\infty}^{LBGR+\Delta} e^{-(x-LBGR)^2/2\sigma^2} dx \\
&= \frac{1}{\sqrt{2\pi}} \int_{-\infty}^{\frac{\Delta}{\sigma}} e^{-x^2/2} dx \\
&= \Phi\left(\frac{\Delta}{\sigma}\right)
\end{aligned}
$$

$$(9\text{-}2)$$

Values of p as a function of Δ/σ, computed from Equation 9-2, can be found in Table 9.3, or in the table of the cumulative normal distribution (Appendix A, Table A.1).

The factor $1/[4(p - 0.5)^2]$ in Equation 9-1 can be viewed as a multiplier applied to the sample sizes given in Table 9.2. If $p = 1$, this factor is one. Figure 9.2 shows the dependence of both p and the sample size multiplier on the resolution of the decision problem, Δ/σ. Increasing Δ/σ beyond about three has little effect on reducing the sample size multiplier. Decreasing Δ/σ below about one causes the sample size multiplier to rise dramatically. The range from the smallest (3) to largest (22) sample size in Table 9.2 is about a factor of seven. When Δ/σ is less than about 0.5, the sample size multiplier exceeds seven. Thus, small values of Δ/σ can have a bigger impact on increasing the sample size than the choice of acceptable decision error rates.

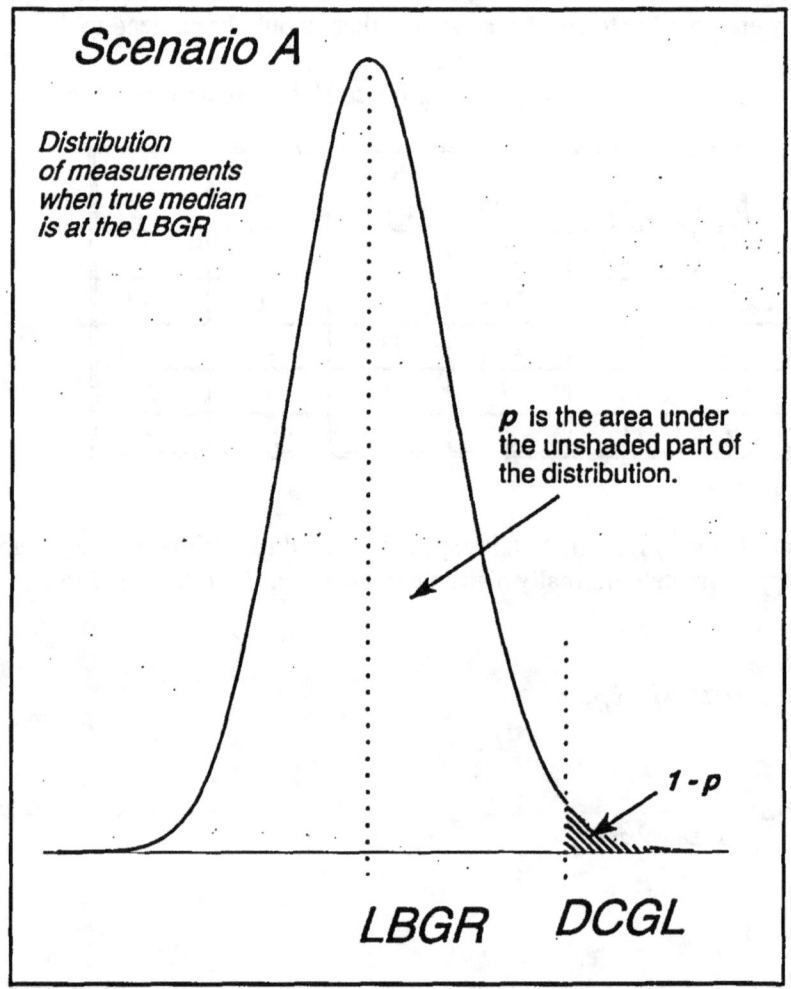

Figure 9.1 The Parameter p for the Sign Test Under Scenario A

The assumption of normality is not critical in the above calculations, since it is only being used to estimate an efficient sample size. However, if a different distribution is considered more appropriate, it can be used. Values of p for other probability distributions with density function $f(x)$, mean equal to the LBGR, and standard deviation σ, can be computed from

$$p = \int_{-\infty}^{DCGL_W} f(x)\,dx \tag{9-3}$$

Table 9.3 Values of p for Use in Computing Sample Size for the Sign Test

Δ/σ	p	Δ/σ	p	Δ/σ	p	Δ/σ	p
0.1	0.53983	1.1	0.86433	2.1	0.98214	3.1	0.99903
0.2	0.57926	1.2	0.88493	2.2	0.9861	3.2	0.99931
0.3	0.61791	1.3	0.9032	2.3	0.98928	3.3	0.99952
0.4	0.65542	1.4	0.91924	2.4	0.9918	3.4	0.99966
0.5	0.69146	1.5	0.93319	2.5	0.99379	3.5	0.99977
0.6	0.72575	1.6	0.9452	2.6	0.99534	4.0	0.99997
0.7	0.75804	1.7	0.95544	2.7	0.99653	5.0	1.00000
0.8	0.78815	1.8	0.96407	2.8	0.99745		
0.9	0.81594	1.9	0.97128	2.9	0.99813		
1.0	0.84135	2.0	0.97725	3.0	0.99865		

In some situations, it may be possible to estimate p directly from remediation control survey data, since it is simply an estimate of the proportion of the final status survey measurements that are expected to fall below the $DCGL_w$.

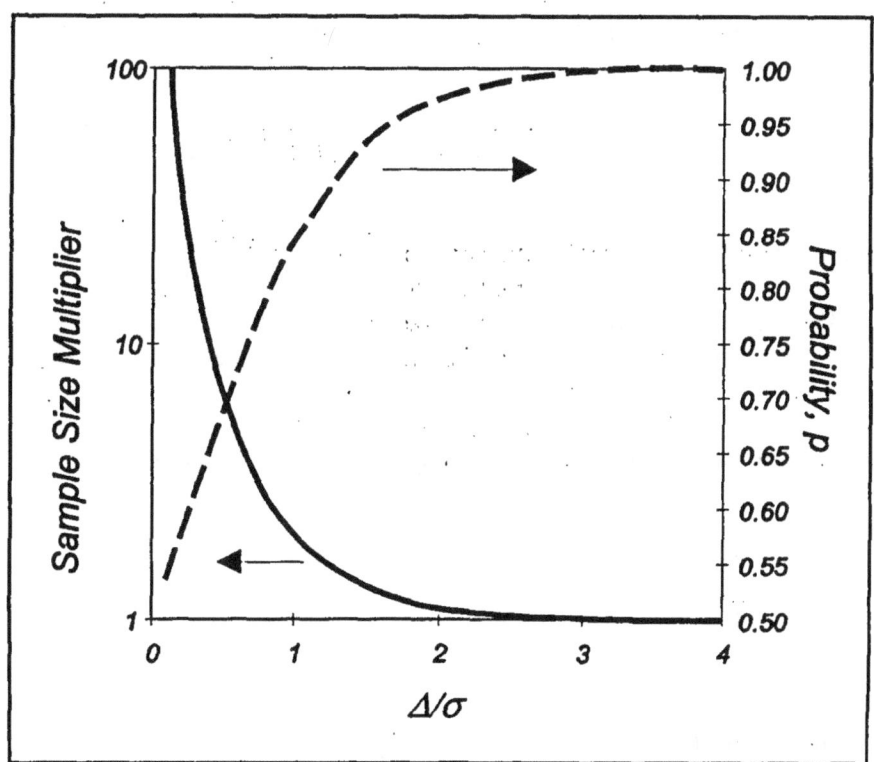

Figure 9.2 Sample Size Multiplier and the Parameter p Versus Δ/σ

NUREG-1505

One can also calculate p from the estimated odds that a random measurement is less than the $DCGL_w$ versus that it is above the $DCGL_w$. If these odds are $r_1 : r_2$, then $p = r_1 / (r_1 + r_2)$. For example, if the odds that a random measurement is less than the $DCGL_w$ are 3:2, then $p = 3/(3+2) = 3/5 = 0.6$.

Whatever method is used to estimate p, it is important not to overestimate it, since that will result in a sample size inadequate to achieve the desired power of the test. The dependence of the sample size multiplier, $1/[4(p - 0.5)^2]$, on p is shown in Figure 9.3.

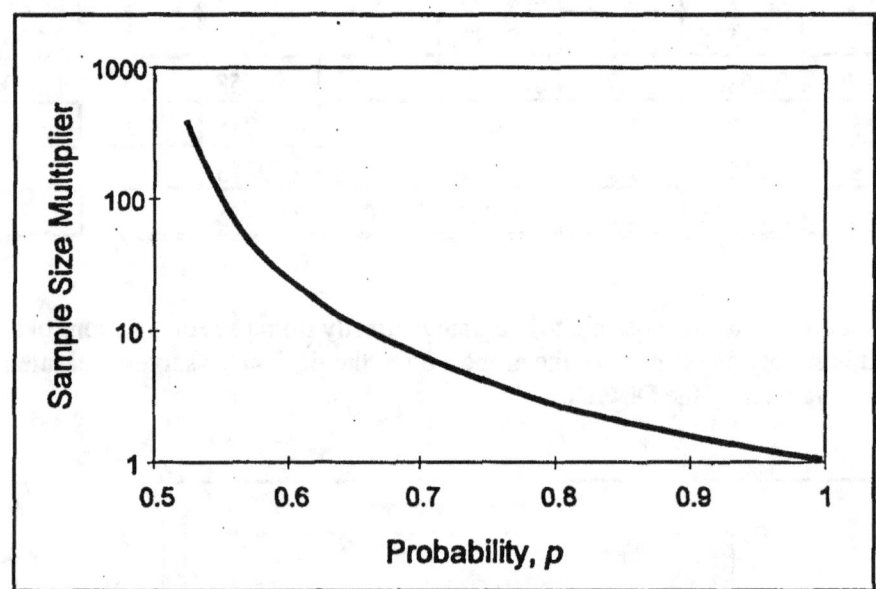

Figure 9.3 Dependence of Sample Size on p

As an illustration, consider the example given in Section 5.1. For that example, the $DCGL_w = 15.9$, the LBGR = 11.5, $\alpha = \beta = 0.05$, and $\sigma = 3.3$. From Table 9.2, $(Z_{1-\alpha} + Z_{1-\beta})^2 = 11$ when $\alpha = \beta = 0.05$. This is the minimum sample size required for those values of the acceptable error rates. The width of the gray region, $\Delta = DCGL_w - LBGR = 15.9 - 11.5 = 4.4$, so $\Delta/\sigma = 4.4/3.3 = 1.3$.

From Table 9.3, the value of p using the normal approximation is 0.903199. Thus the factor

$$
\begin{aligned}
1/[4(p - 0.5)^2] &= 1/[4(0.903199 - 0.5)^2] \\
&= 1/[4(0.403199)^2] \\
&= 1/[4(0.162569)] \\
&= 1/0.650278 \\
&\approx 1.54
\end{aligned}
$$

So, the minimum sample size of 11 is increased by a factor of 1.54 to 16.9. This would normally be rounded up to 17. However, because Equation 9-1 is an approximation, it is prudent to increase this number moderately. An increase of 20% is recommended. This increases the number of samples to $1.2(16.9) = 20.28$, which is rounded up to 21. This is the number that appears in Table 3.2.

The effect of increased variability in the measurement data will be an increase in the required sample sizes. As Δ/σ becomes smaller, p also becomes smaller. This decreases the denominator of Equation 9-1, increasing the sample size N accordingly.

9.3 Sample Size Calculation for the Sign Test Under Scenario B

Under Scenario B, Equation 9-1 is also used to estimate the required sample size. The roles of α and β are reversed, but this has no effect on the numerator of Equation 9-1, so Table 9.2 may still be used. The form of the denominator also remains the same, and Figure 9.3 still represents the dependence of the sample size multiplier on p. However, the definition of the parameter p is different. The definition of the parameter p under Scenario B is the estimated probability that a random measurement from the survey unit will be greater than the LBGR when the survey unit median is actually at the $DCGL_w$. This is illustrated in Figure 9.4. The value of $1-p$ expresses the likelihood that measurements less than the LBGR will be observed, even if half of the concentration distribution is above the $DCGL_w$. This likelihood is higher when the measurement standard deviation is large compared to the width of the gray region.

If, as in Scenario A, we assume that the data are approximately normally distributed, the width of the gray region, $\Delta/\sigma = (DCGL_w - LBGR)/\sigma$, can be used to estimate the parameter p:

$$
\begin{aligned}
p &= \frac{1}{\sqrt{2\pi}\,\sigma} \int_{LBGR}^{\infty} e^{-(x-DCGL_w)^2/2\sigma^2} dx \\[2mm]
&= \frac{1}{\sqrt{2\pi}\,\sigma} \int_{(LBGR-DCGL_w)+DCGL_w}^{-\infty} e^{-(x-DCGL_w)^2/2\sigma^2} dx \\[2mm]
&= \frac{1}{\sqrt{2\pi}} \int_{\frac{LBGR-DCGL_w}{\sigma}}^{\infty} e^{-x^2/2} dx \\[2mm]
&= \frac{1}{\sqrt{2\pi}} \int_{-\infty}^{\frac{DCGL_w-LBGR}{\sigma}} e^{-x^2/2} dx \\[2mm]
&= \Phi\left(\frac{\Delta}{\sigma}\right)
\end{aligned}
\tag{9-4}
$$

This is the same as Equation 9-2. Even though the definition of p has changed, its value as a function of Δ/σ has not changed.

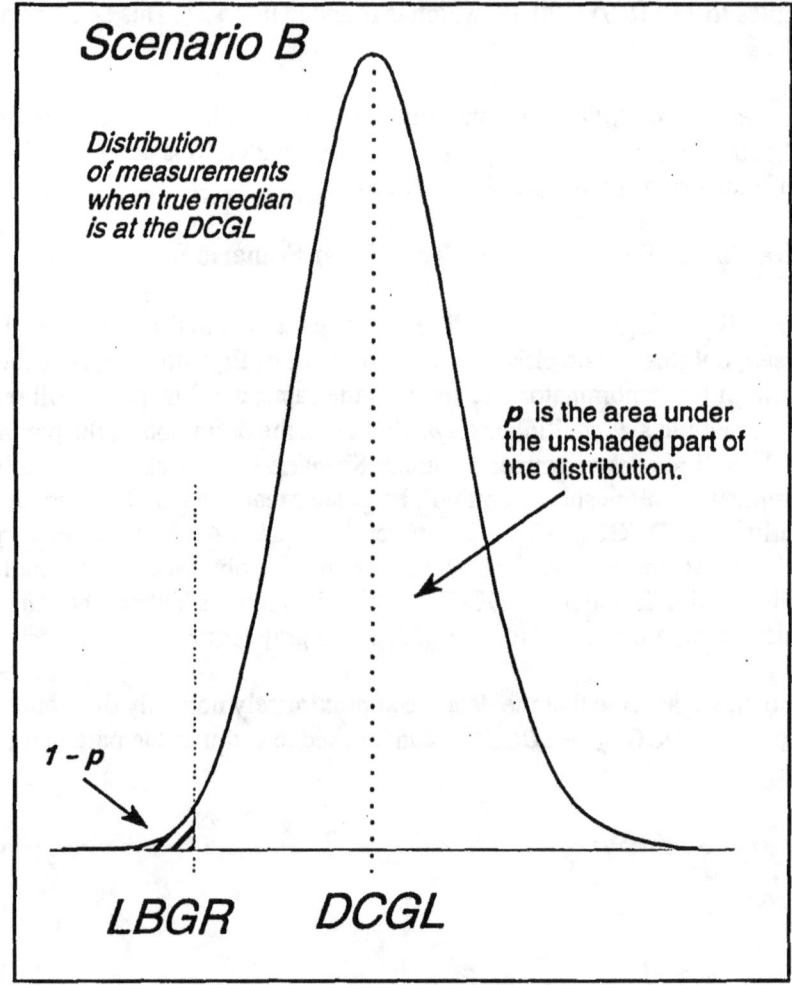

Figure 9.4 The Parameter *p* for the Sign Test Under Scenario B

The above calculation shows that the value of *p* computed from Equation 9-2, as found in Table 9.3, or in the table of the cumulative normal distribution (Appendix A) can be used in Scenario B as well as Scenario A. Figure 9.2, expressing the dependence of sample size on Δ/σ is unchanged, and that is why only one version of Table 3.2 is needed for both scenarios.

If a distribution other than normal is considered more appropriate for determining *p*, the following equation can be used. For a probability distribution with density function $f(x)$, mean at the $DCGL_W$, and standard deviation σ,

$$p = \int_{LBGR}^{\infty} f(x)\,dx \qquad (9\text{-}5)$$

As in Scenario A, it may be possible to estimate p directly from remediation control survey data. To do this, the data should first be shifted so that the median is at the $DCGL_W$ before the proportion that falls above the LBGR is calculated.

p may also be estimated from the odds that a random measurement will be greater than the LBGR versus that it is below the LBGR *when the median is near the $DCGL_W$*. If these odds are $r_1:r_2$, then $p = r_1/(r_1+r_2)$. For example, if the odds that a random measurement is greater than the LBGR when the median is near the $DCGL_W$, are 3:1, then $p = 3/(1+3) = 3/4 = 0.75$. Notice that since we are assuming that the true median is near the $DCGL_W$ the odds must be greater than 1:1. Once a survey unit has been remediated, it may be somewhat unnatural to try to estimate the odds this way.

9.4 Sample Size Calculation for the WRS Test Under Scenario A

For the WRS test, the *total* number of required samples from the reference area and survey unit combined is estimated from (Noether, 1987):

$$N = \frac{(Z_{1-\alpha} + Z_{1-\beta})^2}{12c(1-c)(P_r - 0.5)^2} \tag{9-6}$$

where:
α = specified Type I error rate
β = specified Type II error rate
$Z_{1-\alpha}$ = $100(1-\alpha)$ percentile of the standard normal distribution function
$Z_{1-\beta}$ = $100(1-\beta)$ percentile of the standard normal distribution function
c = proportion of measurements taken in the survey unit.
P_r = estimated probability that a random measurement from the survey unit exceeds a random measurement from the reference area by less than the $DCGL_W$ when the survey unit median is at the LBGR above background. $P_r \neq 0.5$

The numerator of equation 9-6 is the same as that in equation 9-1. Therefore, the same methods are used to calculate it as were discussed in Section 9.2. Table 9.2 gives commonly used values of α and β, together with the corresponding values of $(Z_{1-\alpha} + Z_{1-\beta})^2$. For planning purposes, c is set equal to 0.5, so that Equation 9-6 becomes

$$N = \frac{(Z_{1-\alpha} + Z_{1-\beta})^2}{3(P_r - 0.5)^2} \tag{9-6'}$$

The denominator of Equation 9-6 differs from that in Equation 9-1 in three important ways. First, the parameter P_r replaces the parameter p. It is a different probability, but because it is still a probability, $0 \leq P_r \leq 1$. The factor $(P_r - 0.5)^2$ cannot be larger than 0.25. Second, the constant factor in the denominator is 3 rather than 4. This means that the sample size multiplier $1/[3(P_r - 0.5)^2]$ cannot be smaller than 4/3. Third, Equation 9.6 yields the total number of samples required in both the survey unit and a reference area. $N/2$ samples will be taken in each.

The definition of the parameter P_r states that it is the estimated probability that a random measurement from the survey unit exceeds a random measurement from the reference area by less than the $DCGL_W$ when the survey unit median is above the reference area median by an amount equal to the concentration value at the LBGR. This is illustrated in Figure 9.5.

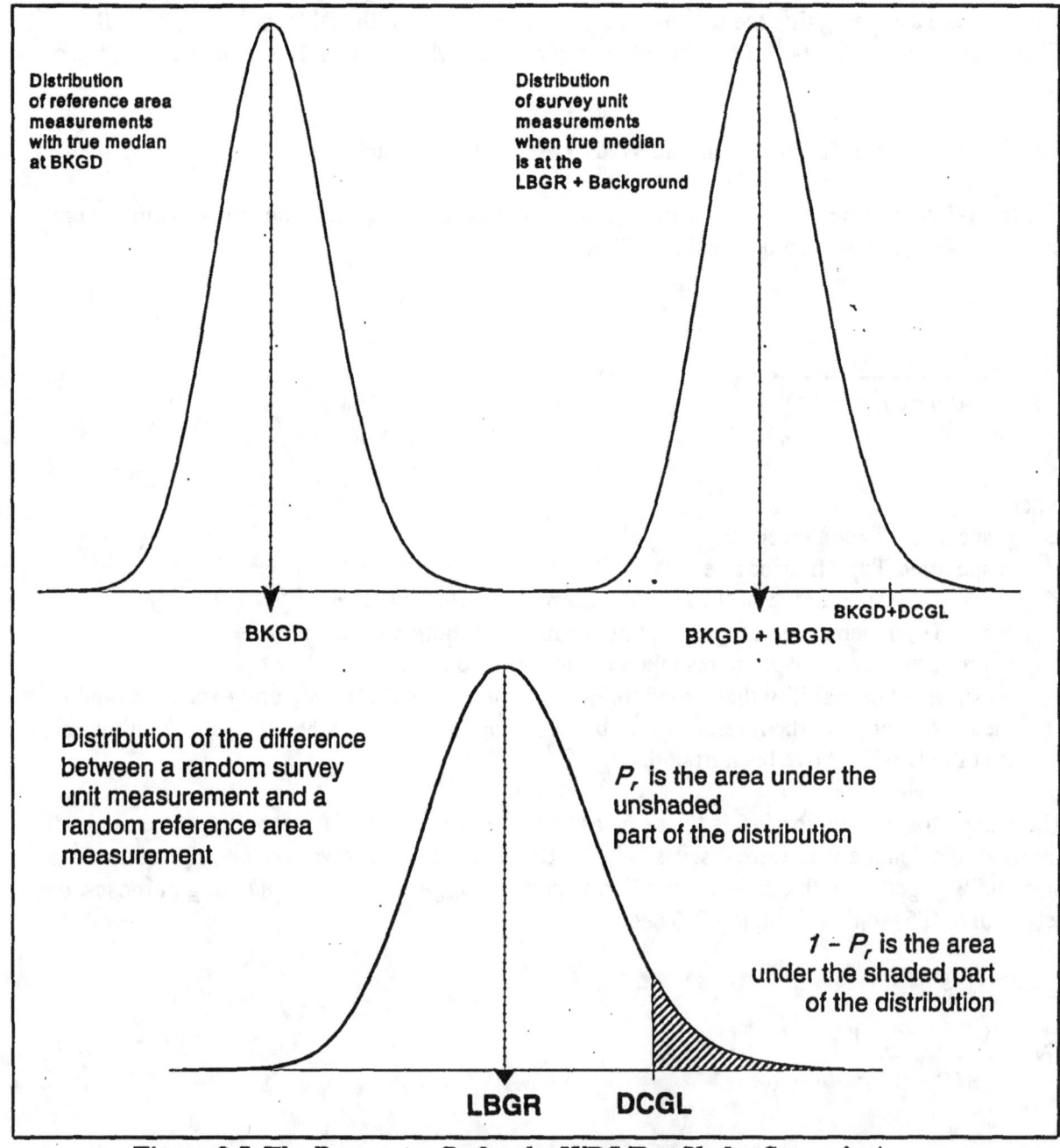

Figure 9.5 The Parameter P_r for the WRS Test Under Scenario A

As was done for the Sign test, the normal distribution may be used to facilitate the conversion of the values of Δ/σ to values of P_r in order to calculate the required sample sizes. The normal distribution is not used to actually conduct the test. Values of P_r are computed for a normal distribution from the following equation:

$$P_r = \text{Probability}(U = X - Y < DCGL)$$

$$= \int_{-\infty}^{DCGL} \left[\int_{-\infty}^{\infty} f_X(u+y) f_Y(y) dy \right] du$$

$$= \int_{-\infty}^{DCGL} \left[\int_{-\infty}^{\infty} \frac{1}{\sqrt{2\pi}\sigma} e^{-(u+y-LBGR-BKGD)^2/2\sigma^2} \frac{1}{\sqrt{2\pi}\sigma} e^{-(y-BKGD)^2/2\sigma^2} dy \right] du$$

$$= \frac{1}{\sqrt{2\pi}\sqrt{2}\sigma} \int_{-\infty}^{DCGL} e^{-(u-LBGR)^2/4\sigma^2} du$$

$$= \frac{1}{\sqrt{2\pi}} \int_{-\infty}^{\frac{DCGL_W - LBGR}{\sqrt{2}\sigma}} e^{-x^2/2} dx$$

$$= \Phi\left(\frac{\Delta}{\sqrt{2}\sigma}\right) \tag{9-7}$$

Values of P_r as a function of Δ/σ are listed in Table 9.4. The probability P_r, and the sample size multiplier $1/[3(P_r - 0.5)^2]$ are shown as a function of Δ/σ in Figure 9.6.

Table 9.4 Values of P_r for Use in Computing Sample Size for the WRS Test

Δ/σ	P_r	Δ/σ	P_r	Δ/σ	P_r	Δ/σ	P_r
0.1	0.528186	1.1	0.781662	2.1	0.931218	3.1	0.985811
0.2	0.556231	1.2	0.801928	2.2	0.940103	3.2	0.988174
0.3	0.583998	1.3	0.821015	2.3	0.948062	3.3	0.990188
0.4	0.611351	1.4	0.838901	2.4	0.955157	3.4	0.991895
0.5	0.638163	1.5	0.855578	2.5	0.961450	3.5	0.993336
0.6	0.664313	1.6	0.871050	2.6	0.967004	4.0	0.997661
0.7	0.689691	1.7	0.885334	2.7	0.971881	5.0	0.999796
0.8	0.714196	1.8	0.898454	2.8	0.976143	6.0	0.999989
0.9	0.737741	1.9	0.910445	2.9	0.979848		
1.0	0.760250	2.0	0.921350	3.0	0.983053		

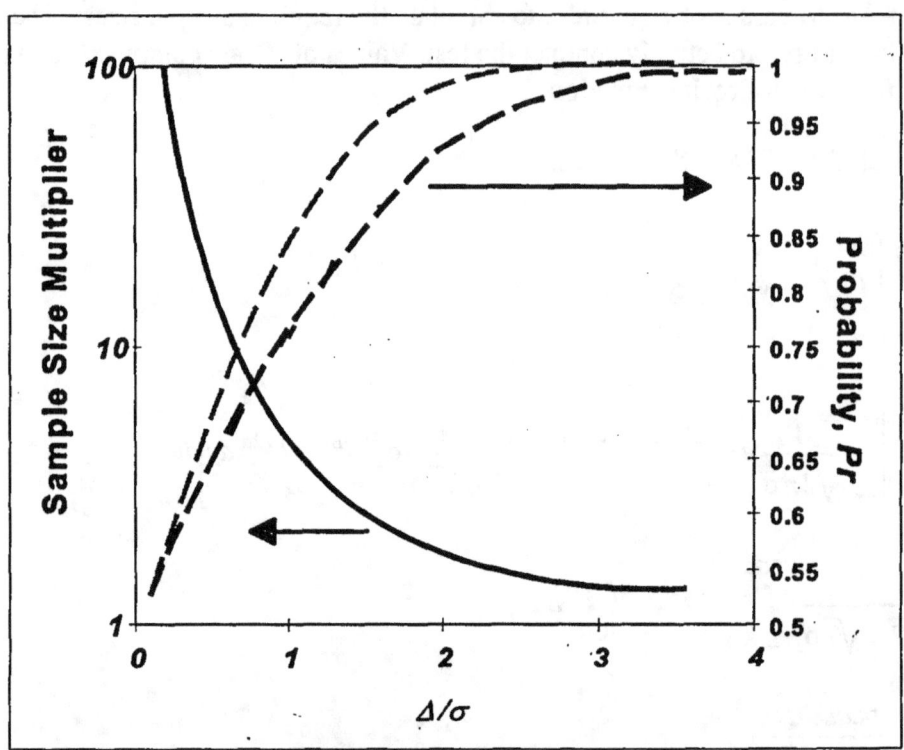

Figure 9.6 Dependence of Sample Size on Δ/σ for WRS Test
(Values of p used for the Sign Test are shown in gray for comparison)

Notice that the only difference between Equation 9-7, and equation 9-4 for computing p for the Sign test is that the standard deviation σ is replaced by $\sqrt{2}$ times σ. This is because the variance of the difference of two independent measurements is the sum of the variances of the individual measurements. If the variances of the individual measurements are about the same, i.e., σ^2, the variance of their difference is $2\sigma^2$. Thus, for a given value of Δ/σ, P_r will always be less than p. This also causes the total sample size required for the WRS test to be greater than that for the Sign test. Values of p are shown for comparison to P_r by the gray dashed line in Figure 9.6.

The combined effect of all the differences between Equation 9-7 and 9-4 is summarized in Figure 9.7.

Values of P_r for distributions other than normal can be calculated from the following equation:

$$P_r = \text{Probability}(U = X - Y < DCGL) = \int_{-\infty}^{DCGL}\left[\int_{-\infty}^{\infty}f_X(u+y)f_Y(y)dy\right]du \qquad (9\text{-}8)$$

where Y is a random measurement from the reference area with density f_Y and X is a random measurement from the survey unit with density f_X. However, in PNL-8989 (1993), Hardin and Gilbert have found that using the values of P_r from Equation 9-6 yielded good results when the distributions being tested were positively skewed, such as the log-normal.

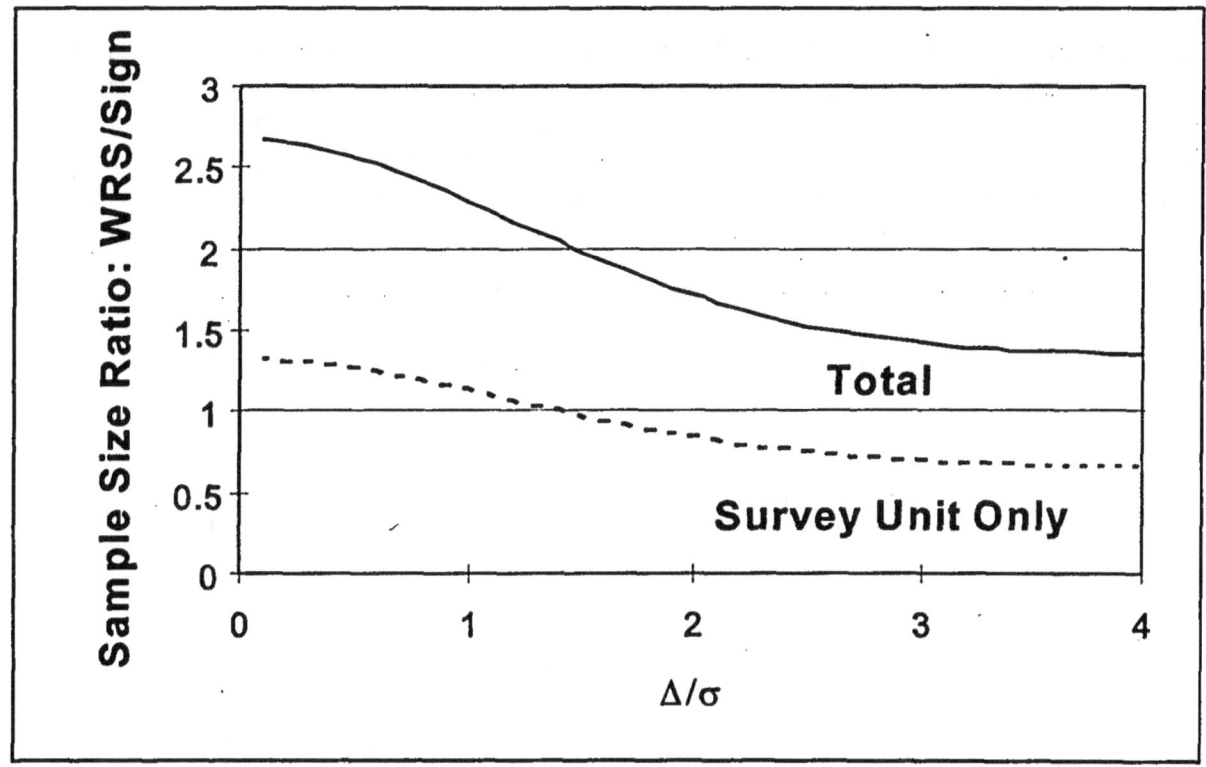

Figure 9.7 Comparison of Sample Sizes Required for the WRS Test and the Sign Test

In some situations, it may be possible to estimate P_r directly from remediation control survey data. It is an estimate of the proportion of time that a random survey measurement will exceed a random reference area measurement by less than the $DCGL_W$.

P_r may also be estimated by the odds that a random survey measurement will exceed a random reference area measurement by less than the $DCGL_W$ versus that a random survey measurement will exceed a random reference area measurement by more than the $DCGL_W$. If these odds are $r_1 : r_2$, then $P_r = r_1 / (r_1 + r_2)$.

Whatever method is used to estimate P_r, it is important not to overestimate it, since that will result in a sample size inadequate to achieve the desired power of the test. The dependence of the sample size multiplier, $1/[3(P_r - 0.5)^2]$, on P_r is shown in Figure 9.8.

As an illustration, consider the example given in Chapter 6.1. For that example, the $DCGL_W = 160$, the $LBGR = 142$, $\alpha = \beta = 0.05$, and $\sigma = 6$. From Table 9.2, $(Z_{1-\alpha} + Z_{1-\beta})^2 = 11$ when $\alpha = \beta = 0.05$. The width of the gray region, $\Delta = DCGL_W - LBGR = 160 - 142 = 18$, so $\Delta/\sigma = 18/6 = 3$.

From Table 9.4, the value of P_r using the normal approximation is 0.983053. Thus the factor $1/[3(p - 0.5)^2] = 1/[3(0.983053 - 0.5)^2]$

$$= 1/[3(0.483053)^2]$$
$$= 1/[3(0.233340)]$$
$$= 1/0.700021$$
$$\approx 1.43.$$

So, the minimum sample size of 11 is increased by a factor of 1.43 to 15.7. This would normally be rounded up to 16, or 8 samples each in the reference area and the survey unit. However, because Equation 9-6 is an approximation, it is prudent to increase this number moderately. An increase of 20% is recommended. This increases the number of samples to $1.2(15.7) = 18.9$, which is rounded up to the next even integer, 20. Thus, 10 samples each in the reference area and the survey unit are required. This is the number that appears in Table 3.3.

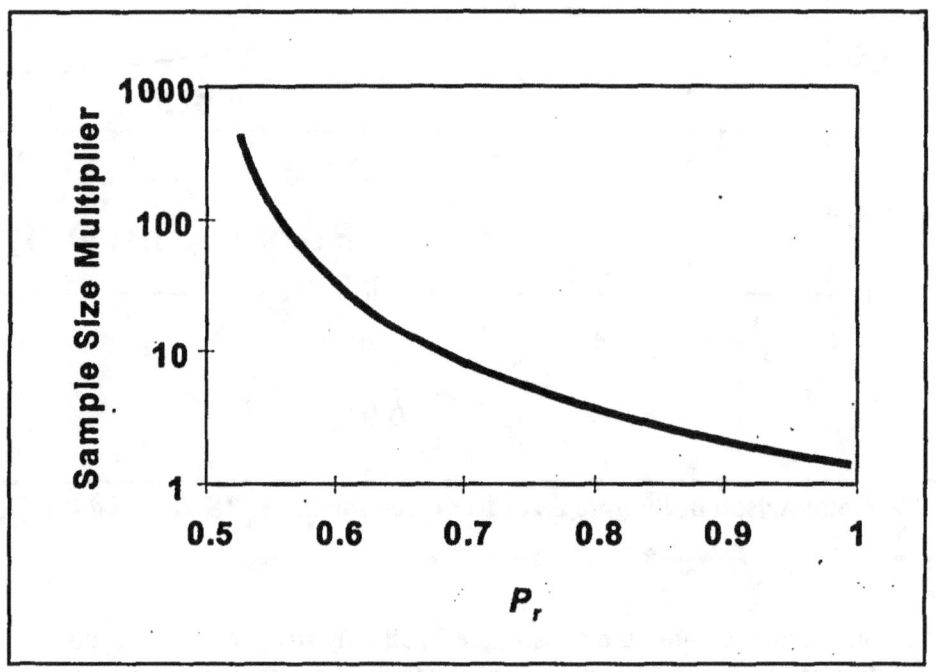

Figure 9.8 Dependence of Sample Size on P_r

9.5 Sample Size Calculation for the WRS Test Under Scenario B

Under Scenario B, Equation 9-6 is also used to estimate the required sample size. The roles of α and β are reversed, but this has no effect on the numerator of Equation 9-6, so Table 9.2 may still be used. However, since under Scenario B, both the WRS test and the Quantile test are used in tandem, the value of α decided on during the DQO process is halved for each test. Thus, the Table 9.2 value for $\alpha_w = \alpha/2$ and β is used.

The form of the denominator also remains the same, and Figure 9.8 still represents the dependence of the sample size multiplier on P_r. However, the definition of the parameter P_r is different.   The definition of the parameter P_r under Scenario B is the estimated probability that the difference between a random measurement from the survey unit and a random measurement from the reference area will be greater than the LBGR when the survey unit median is actually at the $DCGL_w$ above the background median. This is illustrated in Figure 9.9.

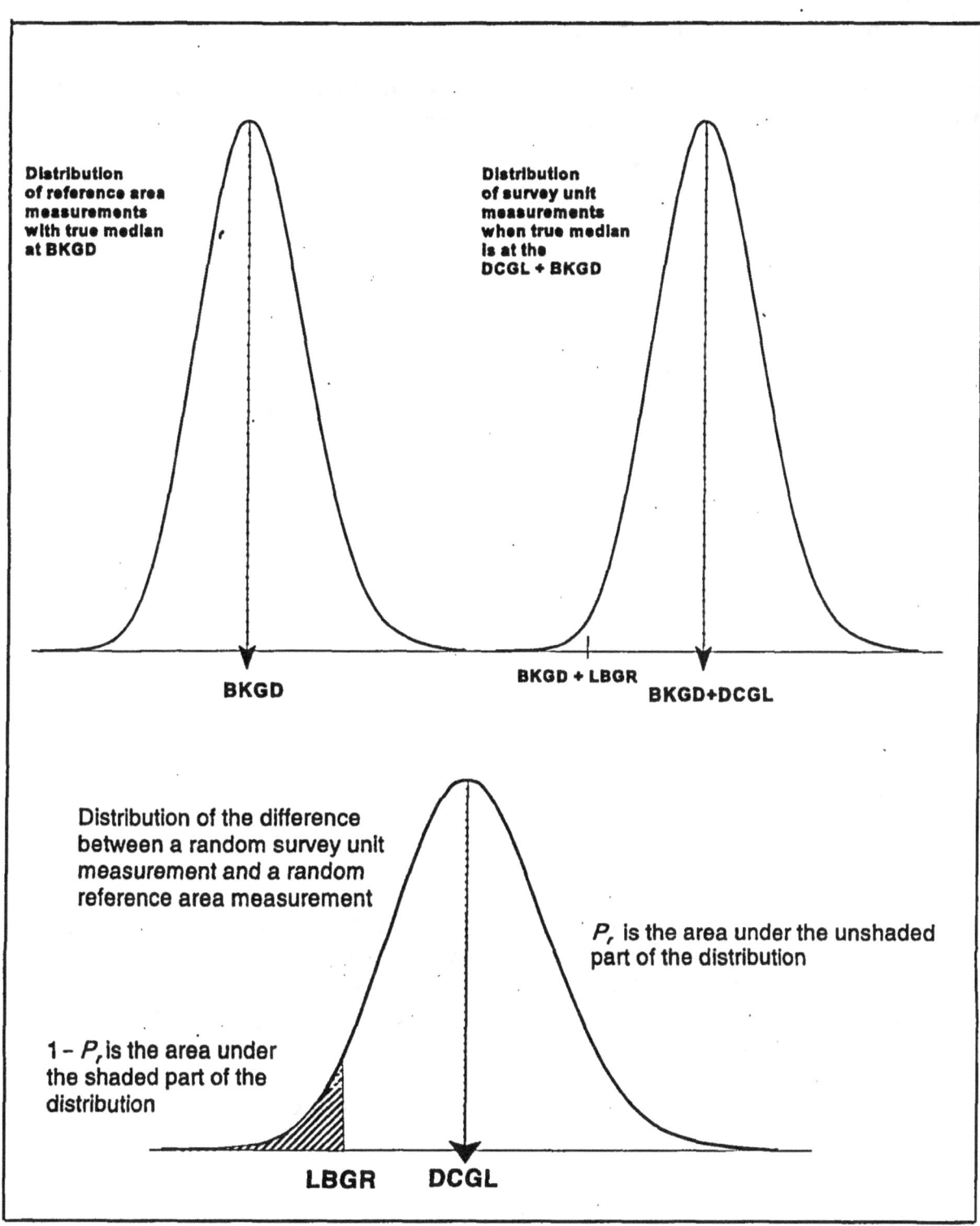

Figure 9.9 The Parameter P_r for the WRS Test Under Scenario B

The value of $1 - P_r$ expresses the likelihood that differences less than the LBGR will be observed, even if half of the survey unit concentration distribution is above the background median by more than the $DCGL_W$. This likelihood is higher when the measurement standard deviation is large compared to the width of the gray region.

If, as in Scenario A, we assume that the data are approximately normally distributed, the width of the gray region, $\Delta/\sigma = (DCGL_W - LBGR)/\sigma$, can be used to estimate the parameter P_r:

$$P_r = \text{Probability}(U = X - Y > LBGR)$$

$$= \int_{LBGR}^{\infty} \left[\int_{-\infty}^{\infty} f_X(u + y) f_Y(y) dy \right] du$$

$$= \int_{LBGR}^{-\infty} \left[\int_{-\infty}^{\infty} \frac{1}{\sqrt{2\pi}\sigma} e^{-(u+y-DCGL_W-BKGD)^2/2\sigma^2} \frac{1}{\sqrt{2\pi}\sigma} e^{-(y-BKGD)^2/2\sigma^2} dy \right] du$$

$$= \frac{1}{\sqrt{2\pi}\,\sqrt{2}\sigma} \int_{LBGR}^{\infty} e^{-(u-DCGL_W)^2/4\sigma^2} du$$

$$= \frac{1}{\sqrt{2\pi}} \int_{-\infty}^{\frac{DCGL_W-LBGR}{\sqrt{2}\sigma}} e^{-x^2/2} dx$$

$$= \Phi\left(\frac{\Delta}{\sqrt{2}\sigma}\right) \tag{9-9}$$

This is the same as Equation 9-7. Even though the definition of P_r has changed, its value as a function of Δ/σ has not changed. Therefore, the value of P_r computed from Equation 9-7, as found in Table 9.4, can be used in Scenario B as well as in Scenario A. Figure 9.6, expressing the dependence of sample size on Δ/σ is unchanged, and that is why only one version of Table 3.3 is needed for both scenarios.

Values of P_r for distributions other than normal can be calculated from the following equation:

$$P_r = \text{Probability}(U = X - Y > LBGR) = \int_{LBGR}^{\infty} \left[\int_{-\infty}^{\infty} f_X(u + y) f_Y(y) dy \right] du \tag{9-10}$$

where Y is a random measurement from the referenece area with density f_Y and X is a random measurement from the survey unit with density f_X.

As in Scenario A, it may be possible to estimate P_r directly from remediation control survey data. To do this, the data should first be shifted so that the median is at the $DCGL_W$ before the proportion that fall above the LBGR is calculated.

P_r may also be estimated from the odds that a random measurement will be greater than the LBGR versus that it is below the LBGR *when the median is near the $DCGL_W$*. If these odds are $r_1{:}r_2$, then $P_r = r_1/(r_1+r_2)$. For example, if the odds that a random measurement is greater than the LBGR when the median is near the $DCGL_W$, are 3:1, then $P_r = 3/(1+3) = 3/4 = 0.75$. Notice that since we are assuming that the true median is near the $DCGL_W$, the odds must be greater than 1:1. However, once a survey unit has been remediated, it may be somewhat unnatural to try to estimate the odds this way.

10 POWER CALCULATIONS FOR THE STATISTICAL TESTS

10.1 Statistical Power and the Probability of Survey Unit Release

The concept of the statistical power of a test was introduced in Section 2.3.2. The use of this concept in optimizing the design of final status surveys was discussed in Section 3.8.1. The power of a statistical test is defined as the probability that the null hypothesis is rejected when it is false. It is $1-\beta$, where β is the Type II error of the test.

The statistical power will have different implications for survey unit release, depending on whether Scenario A or B is used. The same information can be expressed slightly differently. In this report, it is expressed as the probability that the survey unit passes the statistical test, i.e., the result of the test is the decision that the survey unit may be released.

The relationship between this probability and the Type I and Type II errors was given in Table 3.1. Figures 3.9 through 3.12 show this probability as a function of the true residual radioactivity concentration for selected values of α and β over a range of sample sizes. In many cases, it will be sufficient to check the curve in these figures that corresponds most closely to the situation at hand. In the following sections, the assumptions made and the calculations performed in creating these figures are described.

10.2 Power of the Sign Test Under Scenario A

Recall that for the Sign test in Scenario A, the test statistic, S+, was equal to the number of survey unit measurements below the $DCGL_w$. If S+ exceeds the critical value k, then the null hypothesis that the median concentration in the survey unit exceeds the $DCGL_w$ is rejected, i.e., the survey unit passes this test. The probability that any single survey unit measurement falls below the $DCGL_w$ is found from Equation 9-2 or 9-3. The probability that more than k of the N survey unit measurements fall below the $DCGL_w$ is simply the following binomial probability:

$$\sum_{i=k+1}^{N} \binom{N}{i} [p]^i [1-p]^{N-i} = 1 - \sum_{i=0}^{k} \binom{N}{i} [p]^i [1-p]^{N-i} \approx 1 - \Phi\left(\frac{k-Np}{\sqrt{Np(1-p)}}\right) \quad (10\text{-}1)$$

The indicated approximation is generally used when both Np and $N(1-p)$ are five or greater. $\Phi(z)$ is the cumulative standard normal distribution function given in Table A.1.

With p calculated as in Section 9.2, Equation 10-1 yields the probability that the null hypothesis is rejected when the true median of the residual radioactivity concentration in the survey unit is at the LBGR. This is the power of the test at the LBGR.

The probability, $p(C)$, that any single survey unit measurement falls below the $DCGL_w$ when the survey unit median concentration is at any other value, C, can be determined by simply replacing the value of the *LBGR* in Equation 9-2 with the value of C:

$$p(C) = \frac{1}{\sqrt{2\pi}\ \sigma} \int_{-\infty}^{DCGL_W} e^{-(x-C)^2/2\sigma^2} dx$$

$$= \frac{1}{\sqrt{2\pi}\ \sigma} \int_{-\infty}^{C+(DCGL_W - C)} e^{-(x-C)^2/2\sigma^2} dx$$

$$= \frac{1}{\sqrt{2\pi}} \int_{-\infty}^{\frac{(DCGL_W - C)}{\sigma}} e^{-x^2/2} dx$$

$$= \Phi\left(\frac{(DCGL_W - C)}{\sigma}\right) \tag{10-2}$$

Note that if $C = DCGL_W$, $p(C) = 0.5$. The assumption of normality is not critical in the above calculations, since it is only being used to estimate the power. However, if a different distribution is considered more appropriate, Equation 9-3 can be used to calculate $p(C)$.

When the value of $p(C)$ from Equation 10-2 is inserted in Equation 10-1, we obtain the probability that the null hypothesis is rejected at the concentration C. When $C = DCGL_W$, this probability is the probability of a Type I error, α[1]. This calculation can even be performed for values of C greater than the $DCGL_W$. The probability obtained is still the probability that the null hypothesis is rejected, i.e., that the survey unit passes the test.

If the probability that the null hypothesis is rejected (calculated from Equation 10-1) is plotted against the concentration, C, the result is called a power curve. When the power calculation is performed at the design stage, using an estimated value of σ, it is called a prospective power curve. When the calculation is performed after the survey, using the standard deviation of the survey unit measurements as an estimate of σ, it is called a retrospective power curve.

To illustrate the construction of a power curve, consider the example of Chapter 5. The $DCGL_W$ for this example was 15.9 and the LBGR was 11.5. The DQOs for $\alpha = \beta = 0.05$ resulted in a sample size of $N = 21$, using the estimate that $\sigma = 3.3$. From Table A.3, the critical value for the Sign test with $N = 21$ and $\alpha = 0.05$ is $k = 14$. This is all of the information necessary to construct the prospective power curve. To construct the retrospective power curve, we use the standard deviation of the measurement data, 9.5, as the estimate of σ.

The results of these calculations are shown in Table 10.1 and Figure 10.1.

---

[1] The value of α actually obtained from Equation 10-1 should be close to that specified in the DQOs. It may not exactly equal that value when the sample sizes are small, since the critical value, k, can only take integer values.

Table 10.1 Example Power Calculations: Sign Test Scenario A

	Prospective			Retrospective		
C	$(DCGL_W - C)/\sigma$	$p(C)$ (Eq. 10-2)	power (Eq. 10-1)	$(DCGL_W - C)/\sigma$	$p(C)$ (Eq. 10-2)	power (Eq. 10-1)
0	4.82	1.0000	1.000	1.67	0.9525	1.000
5	3.30	0.9995	1.000	1.15	0.8749	0.989
6	3.00	0.9987	1.000	1.04	0.8508	0.972
7	2.70	0.9965	1.000	0.94	0.8264	0.942
8	2.39	0.9916	1.000	0.83	0.7967	0.884
9	2.09	0.9817	1.000	0.73	0.7673	0.802
10	1.79	0.9633	1.000	0.62	0.7324	0.679
11	1.48	0.9306	1.000	0.52	0.6985	0.544
11.5	1.33	0.9082	0.998	0.46	0.6772	0.459
12	1.18	0.8810	0.991	0.41	0.6591	0.390
13	0.88	0.8106	0.914	0.31	0.6217	0.262
14	0.58	0.7190	0.627	0.20	0.5793	0.151
15	0.27	0.6064	0.217	0.09	0.5359	0.076
15.9	0.00	0.5000	0.039	0.00	0.5000	0.039
16	−0.03	0.4880	0.031	−0.01	0.4960	0.036
17	−0.33	0.3707	0.001	−0.12	0.4522	0.014
18	−0.64	0.2611	0.000	−0.22	0.4129	0.005
19	−0.94	0.1736	0.000	−0.33	0.3707	0.001
20	−1.24	0.1075	0.000	−0.43	0.3336	0.000

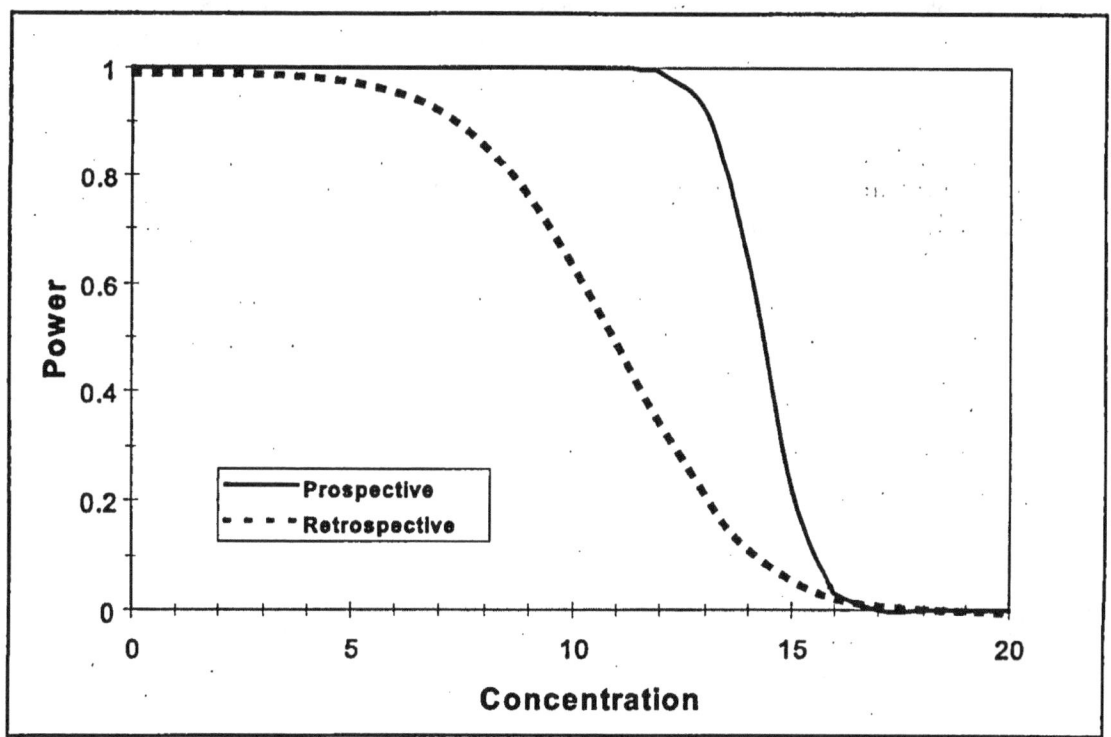

Figure 10.1 Example Power Curves: Sign Test Scenario A

Notice that the increase of σ due to a higher than anticipated measurement standard deviation causes the retrospective power curve to differ considerably from the prospective power curve. In Table 3.3, we see that $\Delta/\sigma = (15.9 - 11.5)/9.5 = 0.46$ results in a much larger required sample size (over 100) to achieve the desired power. Recall that in this example, $S+ = 11$, which is smaller than the critical value $k = 14$. Thus the null hypothesis was not rejected. The survey unit did not pass. We now see that this might have been a consequence of having insufficient power rather than the survey unit actually exceeding the release criterion. The lack of power was due to underestimating the measurement variability.

10.3 Power of the Sign Test Under Scenario B

Recall that for the Sign test in Scenario B, the test statistic, S+, was equal to the number of survey unit measurements above the LBGR. If S+ exceeds the critical value k, then the null hypothesis that the median concentration in the survey unit is less than the LBGR is rejected, i.e. the survey unit does not pass. The probability that any single survey unit measurement falls below the $DCGL_W$, is found from Equation 9-4 or 9-5. The probability that more than k of the N survey unit measurements fall above the LBGR is simply the following binomial probability:

$$\sum_{i=k+1}^{N} \binom{N}{i} [p]^i [1-p]^{N-i} = 1 - \sum_{i=0}^{k} \binom{N}{i} [p]^i [1-p]^{N-i} \approx 1 - \Phi\left(\frac{k-Np}{\sqrt{Np(1-p)}}\right) \qquad (10\text{-}3)$$

The indicated approximation is generally used when both Np and $N(1-p)$ are five or greater. $\Phi(z)$ is the cumulative standard normal distribution function given in Table A.1.

With p calculated as in Section 9.3, this is the probability that the null hypothesis is rejected when the true median of the residual radioactivity concentration in the survey unit is at the $DCGL_W$. This is the power of the test at the $DCGL_W$.

The probability, $p(C)$, that any single survey unit measurement falls above the LBGR when the survey unit median concentration is at any other value, C, can be determined by simply replacing the value of the $DCGL_W$ in Equation 9-4 with the value of C:

$$p(C) = \frac{1}{\sqrt{2\pi}\,\sigma} \int_{LBGR}^{\infty} e^{-(x-C)^2/2\sigma^2} dx$$

$$= \frac{1}{\sqrt{2\pi}\,\sigma} \int_{(LBGR-C)+C}^{-\infty} e^{-(x-C)^2/2\sigma^2} dx$$

$$= \frac{1}{\sqrt{2\pi}} \int_{\frac{LBGR-C}{\sigma}}^{\infty} e^{-x^2/2} dx$$

$$= \frac{1}{\sqrt{2\pi}} \int_{-\infty}^{\frac{C-LBGR}{\sigma}} e^{-x^2/2} dx$$

$$= \Phi\left(\frac{C-LBGR}{\sigma}\right) \tag{10-4}$$

Note that if $C = LBGR$, $p(C) = \Phi(0) = 0.5$. The assumption of normality is not critical in the preceding calculations, since it is only being used to estimate the power. However, if a different distribution is considered more appropriate, Equation 9-5 can be used to calculate $p(C)$.

When the value of $p(C)$ from Equation 10-4 is inserted into Equation 10-2, we obtain the probability that the null hypothesis is rejected at the concentration, C. When $C = LBGR$, this probability is the probability of a Type I error, α[2]. This calculation can even be performed for values of C less than the LBGR. The probability obtained is still the probability that the null hypothesis is rejected, i.e., that the survey unit passes the test, but it is not normally referred to as the power.

If the probability that the null hypothesis is rejected (calculated from Equation 10-3) is plotted against the concentration, C, the result is called a power curve. When the power calculation is performed at the design stage, using an estimated value of σ, it is called a prospective power curve. When the calculation is performed after the survey, using the standard deviation of the survey unit measurements as an estimate of σ, it is called a retrospective power curve.

To illustrate the construction of a power curve, consider the example of Chapter 5. The DCGL$_w$ for this example was 15.9 and the LBGR was 11.5. The DQOs for $\alpha = \beta = 0.05$ resulted in a sample size of $N = 21$, using the estimate that $\sigma = 3.3$. From Table A.3, the critical value for the Sign test with $N = 21$ and $\alpha = 0.05$ is $k = 14$. This is all of the information necessary to construct the prospective power curve. To construct the retrospective power curve, we use the standard deviation of the measurement data, 9.5, as the estimate of σ. The results of these calculations are shown in Table 10.2 and Figure 10.2.

Notice that the increase of σ due to a higher than anticipated measurement standard deviation causes the retrospective power curve to differ considerably from the prospective power curve. $\Delta/\sigma = (15.9 - 11.5)/9.5 = 0.46$ results in a much larger required sample size to achieve the desired power. Recall that in this example, S+ = 13, which is smaller than the critical value, $k = 14$. Thus the null hypothesis was not rejected. The survey unit passes. We now see that this might have been a consequence of having insufficient power rather than the survey unit actually meeting the release criterion. The lack of power was due to underestimating the measurement variability.

---

[2] The value of α actually obtained from Equation 10-2 should be close to that specified in the DQOs. It may not exactly equal that value when the sample sizes are small, since the critical value, k, can only take integer values.

Table 10.2 Example Power Calculations: Sign Test Scenario B

C	Prospective			Retrospective		
	$(C - LBGR)/\sigma$	$p(C)$ (Eq. 10-3)	power (Eq. 10-4)	$(C - LBGR)/\sigma$	$p(C)$ (Eq. 10-2)	power (Eq. 10-1)
0	−3.48	0.0003	0.000	−1.21	0.1131	0.000
5	−1.97	0.0244	0.000	−0.68	0.2483	0.000
6	−1.67	0.0475	0.000	−0.58	0.2810	0.000
7	−1.36	0.0869	0.000	−0.47	0.3192	0.000
8	−1.06	0.1446	0.000	−0.37	0.3557	0.001
9	−0.76	0.2236	0.000	−0.26	0.3974	0.003
10	−0.45	0.3264	0.000	−0.16	0.4364	0.009
11	−0.15	0.4404	0.010	−0.05	0.4801	0.026
11.5	0.00	0.5000	0.039	0.00	0.5000	0.039
12	0.15	0.5596	0.112	0.05	0.5199	0.057
13	0.45	0.6736	0.445	0.16	0.5636	0.119
14	0.76	0.7764	0.830	0.26	0.6026	0.207
15	1.06	0.8554	0.976	0.37	0.6443	0.336
15.9	1.33	0.9082	0.998	0.46	0.6772	0.459
16	1.36	0.9131	0.999	0.47	0.6808	0.474
17	1.67	0.9525	1.000	0.58	0.7190	0.627
18	1.97	0.9756	1.000	0.68	0.7517	0.750
19	2.27	0.9884	1.000	0.79	0.7852	0.854
20	2.58	0.9951	1.000	0.89	0.8133	0.919
21	2.88	0.9980	1.000	1.00	0.8413	0.962
22	3.18	0.9993	1.000	1.11	0.8665	0.984
23	3.48	0.9997	1.000	1.21	0.8869	0.993
24	3.79	0.9999	1.000	1.32	0.9066	0.998
25	4.09	1.0000	1.000	1.42	0.9222	0.999

In Scenario A, the power and the probability that the survey unit passes the test are equivalent. In Scenario B, the power is equivalent to the probability that the survey unit does not pass. To plot the probability that the survey unit passes, the power is subtracted from 1. The result is shown in Figure 10.3.

10.4 Power of the Wilcoxon Rank Sum Test Under Scenario A

Recall that for the Wilcoxon Rank Sum (WRS) test in Scenario A, the test statistic, W_r, was equal to the sum of the ranks of the reference area measurements adjusted for the DCGL$_w$. If W_r exceeds the critical value W_c, then the null hypothesis that the median concentration in the survey unit exceeds that in the reference area by more than the DCGL$_w$ is rejected, i.e., the survey unit passes this test.

The power of the WRS test is very difficult to calculate exactly. However, a good approximation is available (Lehmann and D'Abrera, 1975, $p(C)$ Chapter 2, Section 3, pp. 69–75).

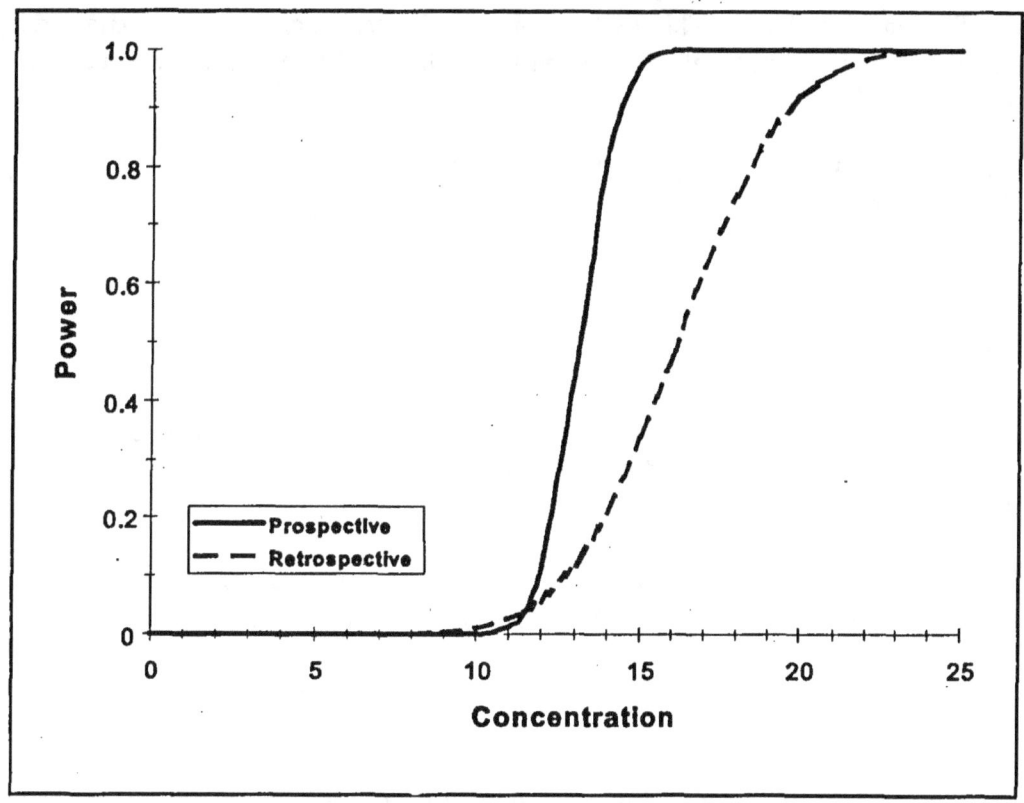

Figure 10.2 Example Power Curves: Sign Test Scenario B

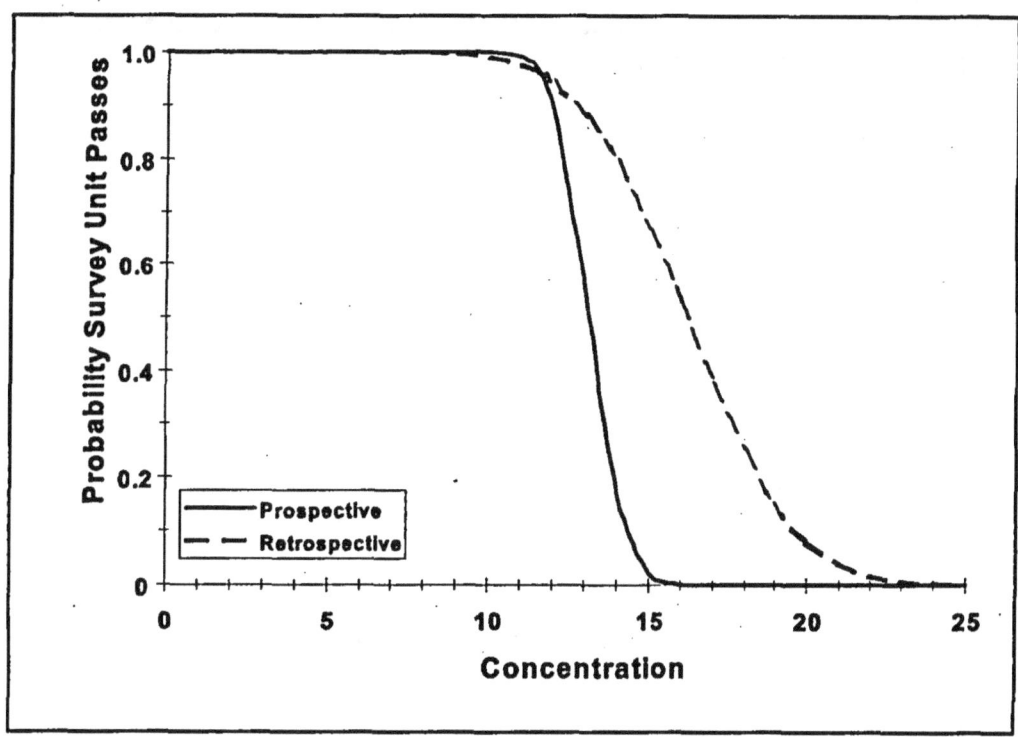

Figure 10.3 Probability Example Survey Unit Passes: Sign Test Scenario B

POWER

If the distribution of the Mann-Whitney form of the WRS test statistic is approximated by a normal distribution, the probability that the null hypothesis will be rejected when the alternative is true can be calculated from:

$$
\text{Power} = 1 - \Phi\left[\frac{W_c - 0.5 - 0.5m(m+1) - E(W_{MW})}{\sqrt{\text{Var}(W_{MW})}}\right]
\tag{10-5}
$$

where W_c is the critical value found in Table A.4 for the appropriate values of the Type I error, α, the number of survey unit measurements, n, and the number of reference area measurements, m. $E(W_{MW})$ and $\text{Var}(W_{MW})$ are the mean and variance of the Mann-Whitney form of the WRS test statistic. Values of $\Phi(z)$, the standard normal cumulative distribution function, are given in Table A.1.

The Mann-Whitney form of the WRS test statistic is $W_{MW} = W_r - 0.5m(m+1)$. It is obtained by subtracting from W_r its minimum value, $0.5m(m+1)$. The mean of W_{MW} is

$$
E(W_{MW}) = mnp_1
\tag{10-6}
$$

where p_1 is the probability that any single measurement from the survey unit exceeds a single measurement from the reference area by less than the $DCGL_W$. This probability depends on the difference in median concentration between the survey unit and the reference area. When this difference is equal to the LBGR, then p_1 is equal to P, as calculated from Equation 9-7. For other values of the difference median concentration between the survey unit and the reference area, C, we simply replace the LBGR in Equation 9-7 with C:

$$
P_r(C) = \text{Probability}(U = X - Y < DCGL)
$$

$$
= \int_{-\infty}^{DCGL}\left[\int_{-\infty}^{\infty} f_X(u+y) f_Y(y)\,dy\right]du
$$

$$
= \int_{-\infty}^{DCGL}\left[\int_{-\infty}^{\infty} \frac{1}{\sqrt{2\pi}\,\sigma} e^{-(u+y-C-BKGD)^2/2\sigma^2}\frac{1}{\sqrt{2\pi}\,\sigma} e^{-(y-BKGD)^2/2\sigma^2}\,dy\right]du
$$

$$
= \frac{1}{\sqrt{2\pi}\,\sqrt{2}\,\sigma}\int_{-\infty}^{DCGL} e^{-(u-C)^2/4\sigma^2}\,du
$$

$$= \frac{1}{\sqrt{2\pi}} \int_{-\infty}^{\frac{DCGL_W - C}{\sqrt{2}\sigma}} e^{-x^2/2} dx$$

$$= \Phi\left(\frac{DCGL_W - C}{\sqrt{2}\sigma} \right) \tag{10-7}$$

Note that if $C = DCGL_W$, then $p_1(C) = 0.5$. The assumption of normality is not critical in the preceding calculations, since it is only being used to estimate the power. However, if a different distribution is considered more appropriate, Equation 9-8 can be used to calculate $p_1(C)$.

The variance of W_{MW} is:

$$\text{Var}(W_{MW}) = mnp_1(1-p_1) + mn(n-1)(p_2 - p_1^2) + mn(m-1)(p_3 - p_1^2) \tag{10-8}$$

p_2 is the probability that two random measurements from the survey unit will each exceed a single random measurement from the reference area by less than the $DCGL_W$; and p_3 is the probability that a single random measurement from the survey unit will exceed each of two random measurements from the reference area unit by less than the $DCGL_W$. When the difference in the concentration distributions of the survey unit and the reference area measurements consists of a shift in the median, and the measurement distributions are symmetric, then $p_2 = p_3$. Then the variance of W_{MW} simplifies to

$$\text{Var}(W_{MW}) = mnp_1(1-p_1) + mn(n+m-2)(p_2 - p_1^2) \tag{10-9}$$

If the measurement distributions are normal, then p_2 is equal to the probability that two correlated standard normal random variables (i.e., with mean = 0 and variance = 1), with correlation coefficient 0.5, are both less than $(DCGL_W - C)/(\sigma\sqrt{2})$. This probability also depends on the difference in median concentration, C, between the survey unit and the reference area. Even with the simplifications employed, the values of p_2 are not easy to calculate. Table 10.3 provides values of p_1 and p_2 as a function of $(DCGL_W - C)/\sigma$ that can be used in calculating the mean and variance of W_{MW}. Nomographs of bivariate normal probabilities that can also be used for this purpose are given in Abramowitz and Stegun (1972).

The power calculated using Equations 10-5 through 10-8 is an approximation. This approximation was compared against the power simulations for the WRS test reported by Gilbert and Simpson (PNL-7409, 1992). It was found that the approximation is sufficiently accurate to determine if the sample design achieves the DQOs.

Table 10.3 Values of p_1 and p_2 for Computing the Mean and Variance of W_{MW} [3]

$(DCGL_W - C)/\sigma$	p_1	p_2	$(DCGL_W - C)/\sigma$	p_1	p_2
-6.0	0.000010	0.000000	0.7	0.689691	0.544073
-5.0	0.000204	0.000010	0.8	0.714196	0.574469
-4.0	0.002339	0.000174	0.9	0.737741	0.604402
-3.5	0.006664	0.000738	1.0	0.760250	0.633702
-3.0	0.016947	0.002690	1.1	0.781662	0.662216
-2.5	0.038550	0.008465	1.2	0.801928	0.689800
-2.0	0.078650	0.023066	1.3	0.821015	0.716331
-1.9	0.089555	0.027714	1.4	0.838901	0.741698
-1.8	0.101546	0.033114	1.5	0.855578	0.765812
-1.7	0.114666	0.039348	1.6	0.871050	0.788602
-1.6	0.128950	0.046501	1.7	0.885334	0.810016
-1.5	0.144422	0.054656	1.8	0.898454	0.830022
-1.4	0.161099	0.063897	1.9	0.910445	0.848605
-1.3	0.178985	0.074301	2.0	0.921350	0.865767
-1.2	0.198072	0.085944	2.1	0.931218	0.881527
-1.1	0.218338	0.098892	2.2	0.940103	0.895917
-1.0	0.239750	0.113202	2.3	0.948062	0.908982
-0.9	0.262259	0.128920	2.4	0.955157	0.920777
-0.8	0.285804	0.146077	2.5	0.961450	0.931365
-0.7	0.310309	0.164691	2.6	0.967004	0.940817
-0.6	0.335687	0.184760	2.7	0.971881	0.949208
-0.5	0.361837	0.206266	2.8	0.976143	0.956616
-0.4	0.388649	0.229172	2.9	0.979848	0.963118
-0.3	0.416002	0.253419	3.0	0.983053	0.968795
-0.2	0.443769	0.278930	3.1	0.985811	0.973725
-0.1	0.471814	0.305606	3.2	0.988174	0.977981
0.0	0.500000	0.333333	3.3	0.990188	0.981636
0.1	0.528186	0.361978	3.4	0.991895	0.984758
0.2	0.556231	0.391392	3.5	0.993336	0.987410
0.3	0.583998	0.421415	4.0	0.997661	0.995497
0.4	0.611351	0.451875	5.0	0.999796	0.999599
0.5	0.638163	0.482593	6.0	0.999989	0.999978
0.6	0.664313	0.513387			

When the values of $p_1(C)$ and $p_2(C)$ and the resulting values of $E(W_{MW})$ and $Var(W_{MW})$ are inserted in Equation 10-5, we obtain the probability that the null hypothesis is rejected at concentration C. When $C = DCGL_W$, this probability is the probability of a Type I error, α. [4]

---

[3] This table may also be used for Scenario B when $(DCGL_W - C)/\sigma$ is replaced by $(C - LBGR)/\sigma$.

[4] The value of α actually obtained from Equation 10-5 should be close to that specified in the DQOs. It may not exactly equal that value when the sample sizes are small, since the critical value, k, can only take integer values.

The preceding calculations can even be performed for values of C greater than the $DCGL_W$. The probability obtained is still the probability that the null hypothesis is rejected, i.e., that the survey unit passes the test.

If the probability that the null hypothesis is rejected (calculated from Equation 10-5) is plotted against the concentration, C, the result is called a power curve. When the power calculation is performed at the design stage, using an estimated value of σ, it is called a prospective power curve. When the calculation is performed after the survey, using the standard deviation of the survey unit measurements as an estimate of σ, it is called a retrospective power curve.

To illustrate the construction of a power curve, consider the example of Chapter 6. The $DCGL_W$ for this example was 160 and the LBGR was 142. The DQOs for $\alpha = \beta = 0.05$ resulted in a sample size of $n = m = 10$, using the estimate that $\sigma = 6$. Twelve samples each were actually taken from the survey unit and the reference area. From Table A.4, the critical value for the WRS test with $n = m = 12$ and $\alpha = 0.05$ is $W_c = 179$. This is all of the information necessary to construct the prospective power curve. To construct the retrospective power curve, we use the larger of the standard deviations of the measurement data from the survey unit and the reference area, 8.1, as the estimate of σ.

The results of these calculations are shown in Table 10.4 and Figure 10.4. In the figure it can be seen that the retrospective power is slightly less than that specified in the DQOs. However, in this example, the null hypothesis was rejected, so the question of the power is moot. The retrospective power calculation is really only necessary when the null hypothesis is not rejected. In that case, it is important to know that it was not rejected simply because there was insufficient power. When the null hypothesis is rejected in spite of insufficient power, the survey designer can consider himself lucky, but the conclusion is still statistically valid.

Table 10.4 Example Prospective Power Calculation: WRS Test Scenario A

C	$(DCGL_W - C)/\sigma$	p_1	p_2	$E(W_{MW})$	$Var(W_{MW})$	$SD(W_{MW})$	z	Power
136	4.00	0.997661	0.995497	143.7	0.9	0.9	-46.21	1.00
139	3.50	0.993336	0.987410	143.0	3.2	1.8	-23.96	1.00
142	3.00	0.983053	0.968795	141.6	10.0	3.2	-12.98	1.00
145	2.50	0.961450	0.931365	138.4	27.4	5.2	-7.24	1.00
148	2.00	0.921350	0.865767	132.7	63.9	8.0	-4.02	1.00
151	1.50	0.855578	0.765812	123.2	124.9	11.2	-2.03	0.98
154	1.00	0.760250	0.633702	109.5	202.8	14.2	-0.63	0.74
157	0.50	0.638163	0.482593	91.9	271.9	16.5	0.52	0.30
160	0.00	0.500000	0.333333	72.0	300.0	17.3	1.65	0.05
163	-0.50	0.361837	0.206266	52.1	271.9	16.5	2.93	0.00
166	-1.00	0.239750	0.113202	34.5	202.8	14.2	4.63	0.00
169	-1.50	0.144422	0.054656	20.8	124.9	11.2	7.13	0.00
172	-2.00	0.078650	0.023066	11.3	63.9	8.0	11.15	0.00

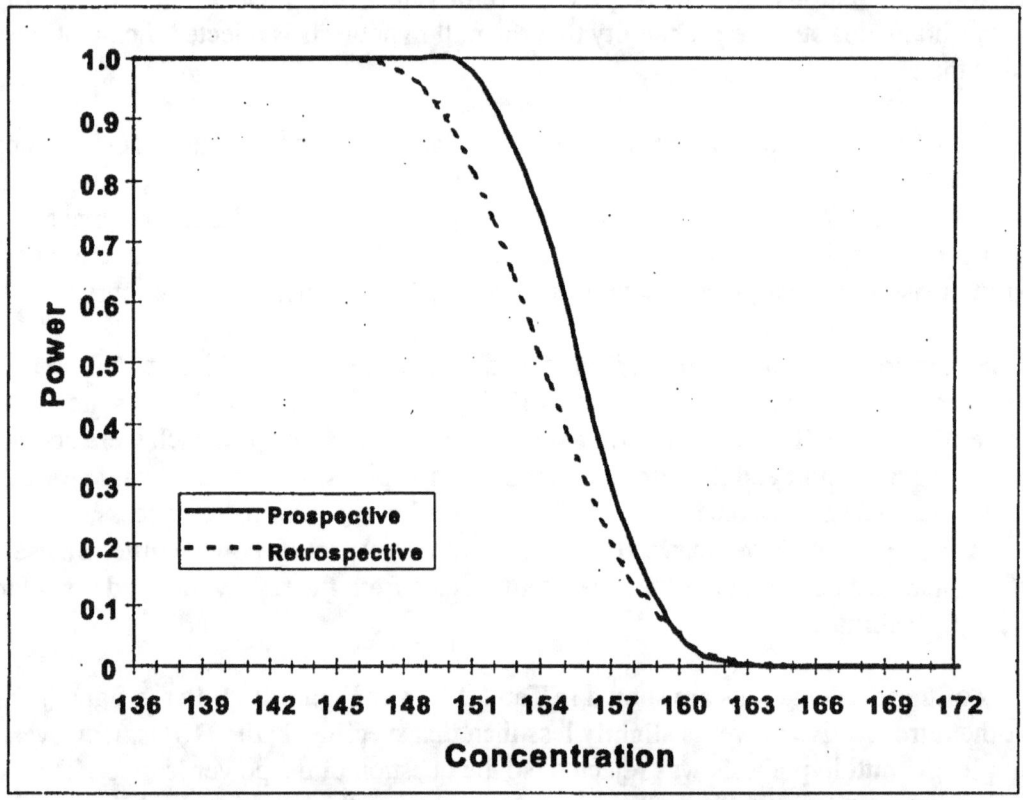

Figure 10.4 Example Power Curves: WRS Test Scenario A

10.5 Power of the Wilcoxon Rank Sum Test Under Scenario B

Recall that for the WRS test in Scenario B, the test statistic, W_s, was equal to the sum of the ranks of the survey unit measurements adjusted for the LBGR. If W_s exceeds the critical value W_c, then the null hypothesis that the median concentration in the survey unit exceeds that in the reference area by less than the LBGR is rejected, i.e., the survey unit does not pass this test.

The power of the WRS test in Scenario B can be approximated in a manner similar to that used in Scenario A, using Equations 10-5, 10-6 and 10-9:

$$\text{Power} = 1 - \Phi\left[\frac{W_c - 0.5 - 0.5m(m+1) - E(W_{MW})}{\sqrt{\text{Var}(W_{MW})}}\right]$$

$$E(W_{MW}) = mnp_1$$

$$\text{Var}(W_{MW}) = mnp_1(1-p_1) + mn(n+m-2)(p_2 - p_1^2)$$

W_c is the critical value found in Table A.4 for the appropriate number of survey unit measurements, n, and number of reference area measurements, m. Since under Scenario B, both the WRS test and the Quantile test are used in tandem, the value of the Type I error, α, decided on during the DQO process, is halved for each test. Thus, the Table A.4 value for value of W_c for $\alpha_W = \alpha/2$ is used. $E(W_{MW})$ and $Var(W_{MW})$ are the mean and variance of the Mann-Whitney form of the WRS test statistic for Scenario B, namely $W_{MW} = W_s - n(n+1)/2$. Values of $\Phi(z)$, the standard normal cumulative distribution function, are given in Table A.1.

In Scenario B, p_1 is the probability that any single measurement from the survey unit exceeds a single measurement from the reference area by more than the LBGR . This probability depends on the difference in median concentration between the survey unit and the reference area. When this difference is equal to the DCGL$_W$, then p_1 is equal to P, as calculated from Equation 9-9. For other values of the difference median concentration between the survey unit and the reference area, C, we simply replace the DCGL$_W$ in Equation 9-9 with C:

$$P_r(C) = \text{Probability}(U = X - Y > LBGR)$$

$$= \int_{LBGR}^{\infty} \left[\int_{-\infty}^{\infty} f_X(u + y) f_Y(y) dy \right] du$$

$$= \int_{LBGR}^{-\infty} \left[\int_{-\infty}^{\infty} \frac{1}{\sqrt{2\pi}\,\sigma} e^{-(u+y-C-BKGD)^2/2\sigma^2} \frac{1}{\sqrt{2\pi}\,\sigma} e^{-(y-BKGD)^2/2\sigma^2} dy \right] du$$

$$= \frac{1}{\sqrt{2\pi}\,\sqrt{2}\,\sigma} \int_{LBGR}^{\infty} e^{-(u-C)^2/4\sigma^2} du$$

$$= \frac{1}{\sqrt{2\pi}} \int_{-\infty}^{\frac{C-LBGR}{\sqrt{2}\sigma}} e^{-x^2/2} dx$$

$$= \Phi\left(\frac{C-LBGR}{\sqrt{2}\sigma} \right) \tag{10-10}$$

This is the same as Equation 10-7, with ($DCGL_W - C$) replaced by ($C - LBGR$). Although the definition of p_1 has changed, its value may still be found from Table 10.3 when ($C - LBGR$)/σ is substituted for ($DCGL_W - C$)/σ. Note that if $C = LBGR$ then $p_1(C) = 0.5$. The assumption of normality is not critical in the above calculations, since it is only being used to estimate the power. However, if a different distribution is considered more appropriate, Equation 9-10 can be used to calculate $p_1(C)$.

In Scenario B, p_2 is the probability that two random measurements from the survey unit will each exceed a single random measurement from the reference area by more than the LBGR ; and p_3 is the probability that a single random measurement from the survey unit will exceed each of two random measurements from the reference area unit by more than the LBGR. When the difference in the concentration distributions of the survey unit and the reference area measurements consists of a shift in the median, and the measurement distributions are symmetric, then $p_2 = p_3$. If the measurement distributions are normal, then p_2 is equal to the probability that two correlated standard normal random variables (i.e., with mean = 0 and variance = 1), with correlation coefficient 0.5, are both less than $(C - LBGR)/(\sigma\sqrt{2})$. This probability also depends on the difference in median concentration, C, between the survey unit and the reference area. Again, values of p_2 may be obtained from Table 10.3 when $(C-LBGR)/\sigma$ is substituted for $(DCGL_W - C)/\sigma$.

Although the power calculated as above is an approximation, this approximation has been compared against the power simulations for the WRS test reported by Gilbert and Simpson (PNL-7409, 1992). It was found that the approximation is sufficiently accurate to determine if the sample design achieves the DQOs.

When the values of $p_1(C)$ and $p_2(C)$ from Table 10.3, and the resulting $E(W_{MW})$ and $Var(W_{MW})$ are inserted in Equation 10-5, we obtain the probability that the null hypothesis is rejected at the concentration C. When $C = DCGL_W$, this probability is the probability of a Type I error, $\alpha_w = \alpha/2$.[5] This calculation can even be performed for values of C less than the LBGR. The probability obtained is still the probability that the null hypothesis is rejected, i.e., that the survey unit passes the test, but it is not usually referred to as the power.

If the probability that the null hypothesis is rejected (calculated from Equation 10-5) is plotted against the concentration, C, the result is called a power curve. When the power calculation is performed at the design stage, using an estimated value of σ, it is called a prospective power curve. When the calculation is performed after the survey, using the standard deviation of the survey unit measurements as an estimate of σ, it is called a retrospective power curve.

To illustrate the construction of a power curve, consider the example of Chapter 6. The $DCGL_W$ for this example was 160 and the LBGR was 142. The DQOs for $\alpha_w = \alpha/2 = 0.025$, and $\beta = 0.05$, result in a sample size of $n = m = 12$, using the estimate that $\sigma = 6$. From Table A.4, the critical value for the WRS test with $n = m = 12$ and $\alpha = 0.025$ is $W_c = 184$. This is all of the information necessary to construct the prospective power curve. To construct the retrospective power curve, we use the larger of the standard deviations of the measurement data from the survey unit and the reference area, 8.1, as the estimate of σ.

The results of these calculations are shown in Table 10.5 and Figure 10.5.

---

[5] The value of α actually obtained from Equation 10-5 should be close to that specified in the DQOs. It may not exactly equal that value when the sample sizes are small, since the critical value, k, can only take integer values.

Table 10.5 Example Prospective Power Calculation: WRS Test Scenario B

C	$(C-LBGR)/\sigma$	p_1	p_2	$E(W_{MW})$	$Var(W_{MW})$	$SD(W_{MW})$	z	Power
136	−1.0	0.239750	0.113202	34.5	202.8	14.2	4.98	0.00
139	−0.5	0.361837	0.206266	52.1	271.9	16.5	3.24	0.00
142	0.0	0.500000	0.333333	72.0	300.0	17.3	1.93	0.03
145	0.5	0.638163	0.482593	91.9	271.9	16.5	0.82	0.20
148	1.0	0.760250	0.633702	109.5	202.8	14.2	−0.28	0.61
151	1.5	0.855578	0.765812	123.2	124.9	11.2	−1.58	0.94
154	2.0	0.921350	0.865767	132.7	63.9	8.0	−3.40	1.00
157	2.5	0.961450	0.931365	138.4	27.4	5.2	−6.29	1.00
160	3.0	0.983053	0.968795	141.6	10.0	3.2	−11.40	1.00
163	3.5	0.993336	0.987410	143.0	3.2	1.8	−21.15	1.00
166	4.0	0.997661	0.995497	143.7	0.9	0.9	−40.85	1.00
172	5.0	0.999796	0.999599	144.0	0.0	0.2	−175.07	1.00

In the figure it can be seen that the retrospective power is slightly less than that specified in the DQOs. However, in this example the null hypothesis was rejected, so the question of the power is moot. The retrospective power calculation is really only necessary when the null hypothesis is accepted. In that case it is important to know that it was not accepted simply because there was insufficient power. When the null hypothesis is rejected in spite of insufficient power, the survey designer can consider himself lucky, but the conclusion is still statistically valid.

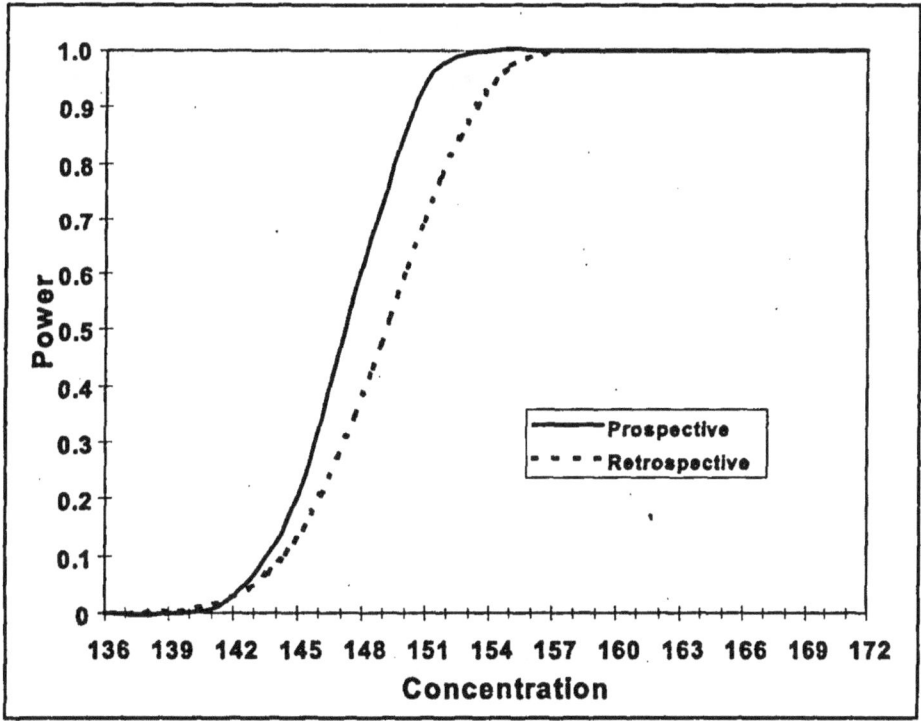

Figure 10.5 Example Power Curves: WRS Test Scenario B

In Scenario A, the power and the probability that the survey unit passes the test are equivalent. In

POWER

Scenario B, the power is equivalent to the probability that the survey unit does not pass. Thus, the probability that the survey unit passes is one minus the power. The result is plotted in Figure 10.6.

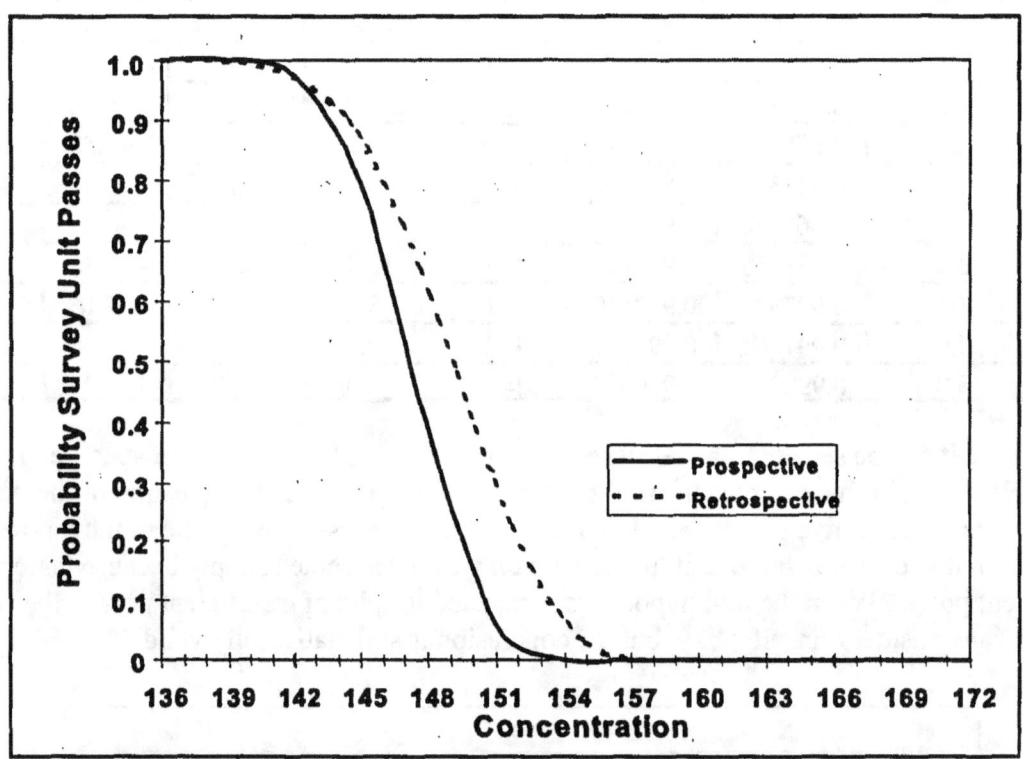

Figure 10.6 Probability Example Survey Unit Passes: WRS Test Scenario B

11 MULTIPLE RADIONUCLIDES

There are two cases to be considered when dealing with multiple radionuclides, namely (1) when the radionuclide concentrations have a fairly constant ratio throughout the survey unit, or (2) when the concentrations of the different radionuclides appear to be unrelated in the survey unit. In statistical terms, we are concerned about whether the concentrations of the different radionuclides are correlated or not. A simple way to judge this would be to make scatter plots of the concentrations against each other, and see if the points appear to have an underlying linear pattern. The correlation coefficient can also be computed to see if it lies nearer to zero than to one. One could also perform a curve fit and test the significance of the result. Ultimately, however, sound judgment must be used in interpreting the results of such calculations. If there is no physical reason for the concentrations to be related, they probably are not. Conversely, if there is sound evidence that the radionuclide concentrations should be related because of how they were treated, processed or released, this information should be used.

11.1 Using the Unity Rule

In either of the two above cases, the unity rule described in Section 3.3 is applied. The difference is in how it is applied. Suppose there are n radionuclides. If the concentration of radionuclide i is denoted by C_i, and its $DCGL_w$ is denoted by D_i, then the unity rule for the n radionuclides states that

$$C_1/D_1 + C_2/D_2 + C_3/D_3 + \cdots + C_n/D_n \leq 1 \qquad (11\text{-}1)$$

This will ensure that the total dose due to the sum of all the radionuclides does not exceed the release criterion. Note that if D_{min} is the smallest of the DCGLs, then

$$(C_1 + C_2 + C_3 + \cdots + C_n)/D_{min} \leq C_1/D_1 + C_2/D_2 + C_3/D_3 + \cdots + C_n/D_n \qquad (11\text{-}2)$$

So that the smallest DCGL may be applied to the total activity concentration, rather than using the unity rule. While it is an option to consider, in many cases this approach will be too conservative to be useful.

11.2 Radionuclide Concentrations With Fixed Ratios

If there is an established ratio among the concentrations of the n radionuclides in a survey unit, then the concentration of every radionuclide can be expressed in terms of any one of them, e.g., radionuclide #1. The measured radionuclide is often called a *surrogate* radionuclide for the others.

If $C_2 = R_2 C_1$, $C_3 = R_3 C_1$, ..., $C_i = R_i C_1$, ..., $C_n = R_n C_1$,
then
$$
\begin{aligned}
C_1/D_1 + C_2/D_2 + C_3/D_3 + \cdots + C_n/D_n &= C_1/D_1 + R_2 C_1/D_2 + R_3 C_1/D_3 + \cdots + R_n C_1/D_n \\
&= C_1 [1/D_1 + R_2/D_2 + R_3/D_3 + \cdots + R_n/D_n] \\
&= C_1/D_{total}, \qquad (11\text{-}3)
\end{aligned}
$$

where

$$D_{total} = 1/ [1/D_1 + R_2/D_2 + R_3/D_3 + \cdots + R_n/D_n]. \tag{11-4}$$

Thus, D_{total} is the $DCGL_W$ for the surrogate radionuclide when the concentration of that radionuclide represents all radionuclides that are present in the survey unit. Clearly, this scheme is applicable only when radionuclide-specific measurements of the surrogate radionuclide are made. It is unlikely to apply in situations where the surrogate radionuclide appears in background, since background variations would tend to obscure the relationships between it and the other radionuclides.

Thus, in the case in which there are constant ratios among radionuclide concentrations, the statistical tests are applied as if only the surrogate radionuclide were contributing to the residual radioactivity, with the $DCGL_W$ for that radionuclide replaced by D_{total}. For example, in planning the final status survey, only the expected standard deviation of the concentration measurements for the surrogate radionuclide is needed to calculate the sample size.

For the elevated measurement comparison, the $DCGL_{EMC}$ for the surrogate radionuclide is replaced by

$$E_{total} = 1/ [1/E_1 + R_2/E_2 + R_3/E_3 + \cdots + R_n/E_n], \tag{11-5}$$

where E_i is the $DCGL_{EMC}$ for radionuclide i.

11.3 Unrelated Radionuclide Concentrations

If the concentrations of the different radionuclides appear to be unrelated in the survey unit, then there is little alternative to measuring the concentration of each radionuclide and using the unity rule. The exception would be in applying the most restrictive $DCGL_W$ to all of the radionuclides, as mentioned in Section 11.1.

Since the release criterion is

$$C_1/D_1 + C_2/D_2 + C_3/D_3 + \cdots + C_n/D_n \le 1.0 \tag{11-6}$$

the quantity to be measured is the *weighted sum*, $T = C_1/D_1 + C_2/D_2 + C_3/D_3 + \cdots + C_n/D_n$. The $DCGL_W$ for T is 1.0. In planning the final status survey, the measurement standard deviation of the weighted sum, T, is estimated by

$$\sigma^2(T) = [\sigma(C_1)/D_1]^2 + [\sigma(C_2)/D_2]^2 + [\sigma(C_3)/D_3]^2 + \cdots + [\sigma(C_n)/D_n]^2, \tag{11-7}$$

since the measured concentrations of the various radionuclides are assumed to be uncorrelated.

For the elevated measurement comparison, the inequality

$$C_1/E_1 + C_2/E_2 + C_3/E_3 + \cdots + C_n/E_n \le 1.0 \tag{11-8}$$

is used, where E_i is the $DCGL_{EMC}$ for radionuclide i. For scanning, most restrictive $DCGL_{EMC}$ should generally be used.

When some of the radionuclides also appear in background, the quantity
$$T = C_1 / D_1 + C_2 / D_2 + C_3 / D_3 + \cdots + C_n / D_n$$
must also be measured in an appropriate reference area. If radionuclide i does not appear in background, set $C_i = 0$ in the calculation of T for the reference area.

Note that if there is a fixed ratio between the concentrations of some radionuclides, but not others, a combination of the method of this section with that of the previous section may be used, using the appropriate value of D_{total} with the concentration of the measured surrogate radionuclide to replace the corresponding terms in Equation 11-7.

11.4 Example Application of WRS Test to Multiple Radionuclides

This section contains an example application of the nonparametric statistical methods in this report to sites that have residual radioactivity from more than one radionuclide. Consider a site with both ^{60}Co and ^{137}Cs contamination. ^{137}Cs appears in background from global atmospheric weapons tests at a typical concentration of about 1 pCi/g. Assume that the $DCGL_W$ for ^{60}Co is 2 pCi/g and that for ^{137}Cs is 1.4 pCi/g. In disturbed areas, the background concentration of ^{137}Cs can vary considerably. An estimated spatial standard deviation of 0.5 pCi/g for ^{137}Cs will be assumed. During remediation it was found that the concentrations of the two radionuclides were not well correlated in the survey unit. ^{60}Co concentrations were more variable than the ^{137}Cs concentrations, and 0.7 pCi/g is assumed for its standard deviation. Measurement errors for both ^{60}Co and ^{137}Cs using gamma spectrometry will be small compared to this. For the comparison to the release criteria, the weighted sum of the concentrations of these radionuclides is computed from the following:

Weighted sum = $(^{60}$Co concentration$)/(^{60}$Co $DCGL_W) + (^{137}$Cs concentration$)/(^{137}$Cs $DCGL_W)$

$\qquad = (^{60}$Co concentration$)/(2) + (^{137}$Cs concentration$)/(1.4)$

The variance of the weighted sum, assuming that the ^{60}Co and ^{137}Cs concentrations are spatially unrelated is

$\sigma^2 = [(^{60}$Co standard deviation$)/(^{60}$Co $DCGL_W)]^2 + [(^{137}$Cs standard deviation$)/(^{137}$Cs $DCGL_W)]^2$

$\qquad = [(0.7)/(2)]^2 + [(0.5)/(1.4)]^2 = 0.25.$

Thus $\sigma = 0.5$. The $DCGL_W$ for the weighted sum is one. Scenario A will be used, i.e., the null hypothesis is that the survey unit exceeds the release criterion. During the DQO process, the LBGR was set at 0.5 for the weighted sum, so that $\Delta = DCGL_W - LBGR = 1.0 - 0.5 = 0.5$, and $\Delta/\sigma = 0.5/0.5 = 1.0$. The acceptable error rates chosen were $\alpha = \beta = 0.05$. To achieve this, 32 samples each are required in the survey unit and the reference area.

The weighted sums are computed for each measurement location in both the reference area and the survey unit. The WRS test is then performed on the weighted sum. The calculations for this example are shown in Table 11.1.

Table 11.1 Example WRS Test for Two Radionuclides

	Reference Area		Survey Unit		Weighted Sum			Ranks	
	^{137}Cs	^{60}Co	^{137}Cs	^{60}Co	Ref	Survey	Adj Ref	Survey	Adj Ref
1	2	0	1.12	0.06	1.43	0.83	2.43	1	56
2	1.23	0	1.66	1.99	0.88	2.18	1.88	43	21
3	0.99	0	3.02	0.56	0.71	2.44	1.71	57	14
4	1.98	0	2.47	0.26	1.41	1.89	2.41	23	55
5	1.78	0	2.08	0.21	1.27	1.59	2.27	9	50
6	1.93	0	2.96	0.00	1.38	2.11	2.38	37	54
7	1.73	0	2.05	0.20	1.23	1.56	2.23	7	46
8	1.83	0	2.41	0.00	1.30	1.72	2.30	16	52
9	1.27	0	1.74	0.00	0.91	1.24	1.91	2	24
10	0.74	0	2.65	0.16	0.53	1.97	1.53	27	6
11	1.17	0	1.92	0.63	0.83	1.68	1.83	13	18
12	1.51	0	1.91	0.69	1.08	1.71	2.08	15	32
13	2.25	0	3.06	0.13	1.61	2.25	2.61	47	63
14	1.36	0	2.18	0.98	0.97	2.05	1.97	30	28
15	2.05	0	2.08	1.26	1.46	2.12	2.46	39	58
16	1.61	0	2.30	1.16	1.15	2.22	2.15	45	41
17	1.29	0	2.20	0.00	0.92	1.57	1.92	8	25
18	1.55	0	3.11	0.50	1.11	2.47	2.11	59	35
19	1.82	0	2.31	0.00	1.30	1.65	2.30	11	51
20	1.17	0	2.82	0.41	0.84	2.22	1.84	44	19
21	1.76	0	1.81	1.18	1.26	1.88	2.26	22	48
22	2.21	0	2.71	0.17	1.58	2.02	2.58	29	62
23	2.35	0	1.89	0.00	1.68	1.35	2.68	3	64
24	1.51	0	2.12	0.34	1.08	1.68	2.08	12	33
25	0.66	0	2.59	0.14	0.47	1.92	1.47	26	5
26	1.56	0	1.75	0.71	1.12	1.60	2.12	10	38
27	1.93	0	2.35	0.85	1.38	2.10	2.38	34	53
28	2.15	0	2.28	0.87	1.54	2.06	2.54	31	61
29	2.07	0	2.56	0.56	1.48	2.11	2.48	36	60
30	1.77	0	2.50	0.00	1.27	1.78	2.27	17	49
31	1.19	0	1.79	0.30	0.85	1.43	1.85	4	20
32	1.57	0	2.55	0.70	1.12	2.17	2.12	42	40
Avg	1.62	0	2.28	0.47	1.16	1.86	2.16	Sum = 799	Sum = 1281
Std Dev	0.43	0	0.46	0.48	0.31	0.36	0.31		

In Scenario A, the $DCGL_W$ (i.e., 1.0) is added to the weighted sum for each location in the reference area. The ranks of the combined survey unit and adjusted reference area weighted sums are then computed. The sum of the ranks of the adjusted reference area weighted sums is then compared to the critical value for $n = m = 32$, $\alpha = 0.05$, which is 1162 (see formula following

Table A.4). In Table 11.1, the sum of the ranks of the adjusted reference area weighted sums is 1281. This exceeds the critical value, so the null hypothesis is rejected. The survey unit meets the release criterion. The difference between the mean of the weighted sums in the survey unit and the reference area is $1.86 - 1.16 = 0.7$. Thus, the estimated dose due to residual radioactivity in the survey unit is 70% of the release criterion.

12 MULTIPLE SURFACES

12.1 Choosing Survey Units

Three criteria mentioned in Section 2.2 constrained the choice of survey units:

(1) Classification by Contamination Potential: Survey units are composed of areas with similar usage, contamination, and remediation histories that determine the requirements of the final status survey.

(2) Congruity With the Dose Model Used: When the release criteria are dose-based, the survey unit configuration should be consistent with those assumed in the dose model used.

(3) Data Variability: The measurement data variability, σ, within survey units should be minimized so that acceptable decision error rates can be obtained with efficient sample sizes.

These criteria should guide the selection of survey units during the DQO process. However, there are situations in which it will necessary to balance the requirements of one criterion against the requirements of another. In those circumstances, one should be guided by the ultimate objective of the final status survey, namely to make the correct decision on whether the survey unit meets the release criterion.

As an example, consider a room with a concrete floor, one wall with tile, another with wallboard, a third with glass doors and windows, and a fourth with a large blackboard. There are at least five different surfaces, with potentially five different levels of residual radioactivity. Using only criteria (1) and (3), it might seem important to treat these as five distinct survey units. However, this is not only very inefficient, it may not even be the best solution. It is unlikely that any dose model treats a contaminated blackboard by itself. Modeling the five areas separately and combining the results may not be as faithful a representation as treating all the surfaces together as one contaminated room.

In this chapter, we discuss some of the factors to be considered during the DQO process to optimize the choice of survey units. We also introduce the concept of performing a *Sign test on paired measurements* when there are many diverse background materials present in a survey unit.

12.2 Combining Dissimilar Areas Into One Survey Unit

The primary disadvantages in separating very small areas of dissimilar contamination potential into distinct survey units are that such small survey units will not generally conform well to dose models, and that the resulting sample densities may be unreasonably high. The disadvantage of combining such areas is that the resulting survey unit will have a larger spatial concentration variability, requiring a larger sample size than if it were more uniform. However, the total sample size may not be as large as would be needed for separate survey units. This possibility can be investigated and resolved during the DQO process.

Consider an area such as shown in Figure 12.1: a 25 m by 100 m gravel parking lot, with a paved walkway across a 25 m by 100 m lawn to a building. The walkway covers about 400 m^2 of the lawn area. There are many ways that this area could be divided into survey units, depending on the level of expected contamination.

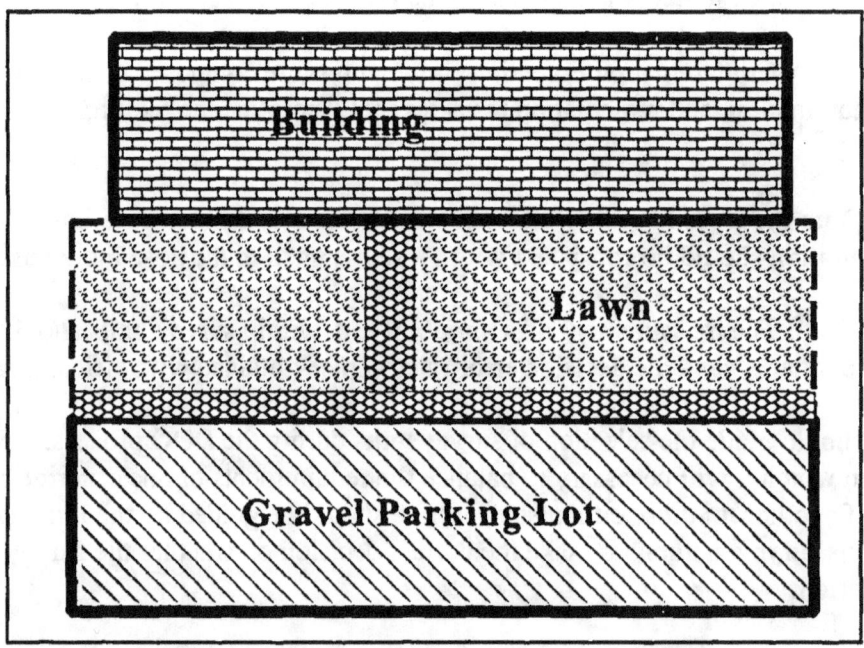

Figure 12.1 Example Survey Units: Case #1

Example Case #1: ^{60}Co was handled in hoods that were vented from the roof of the building. There is the possibility of a small amount of residual radioactivity, but at levels anticipated to be far below the DCGL. There is virtually no possibility of isolated elevated areas. All of the areas would be considered Class 2. Surface samples will be analyzed for ^{60}Co by Ge gamma spectrometry. The total area (5000 m^2) is within the parameters used by the dose model. Although it is possible that average residual radioactivity levels between the walk, the gravel, and the lawn are different, there is probably no reason to divide these areas into separate survey units. The walk may have almost no contamination because of runoff. Judgmental scans should probably be made along the edges of the walk and along the side of the building.

Example Case #2: The same as example 1, except that there had been a spill in the gravel parking lot. The area of the spill is shown as the dark area in Figure 12.2. During remediation of the spill, the surface material of the parking lot was disturbed by earthmoving equipment in the crosshatched area, approximately 1700 m^2. In this case, the disturbed area around the spill should be considered a Class 1 area. The surrounding area is still Class 2. There is the possibility of designating the entire parking lot as a Class 1 survey unit, but it might actually be more reasonable to include the uncontaminated part of the parking lot along with the other Class 2 areas. This will yield a higher sampling density in the actual contaminated area, even though it increases the variability in the Class 2 survey unit.

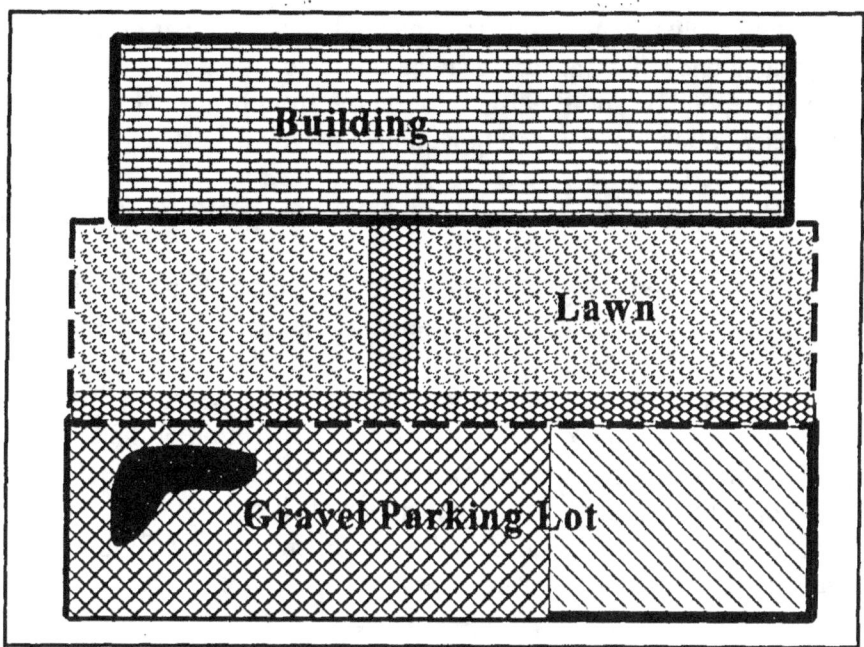

Figure 12.2 Example Survey Units: Case #2

Example Case #3: The same as example 2, except that the spill was on the walkway, as shown in Figure 12.3. The entire paved walkway was removed during remediation, and a substantial portion of the lawn and parking lot were disturbed in the process. The entire area (5000 m^2) should probably be designated Class 1, but is too large to contain only one survey unit. In this case, it may be reasonable to divide the area into two survey units— the former lawn area and the former parking lot. It is probably not practical, or prudent, to try to separate the small undisturbed parts of the lawn and the parking lot into a third survey unit.

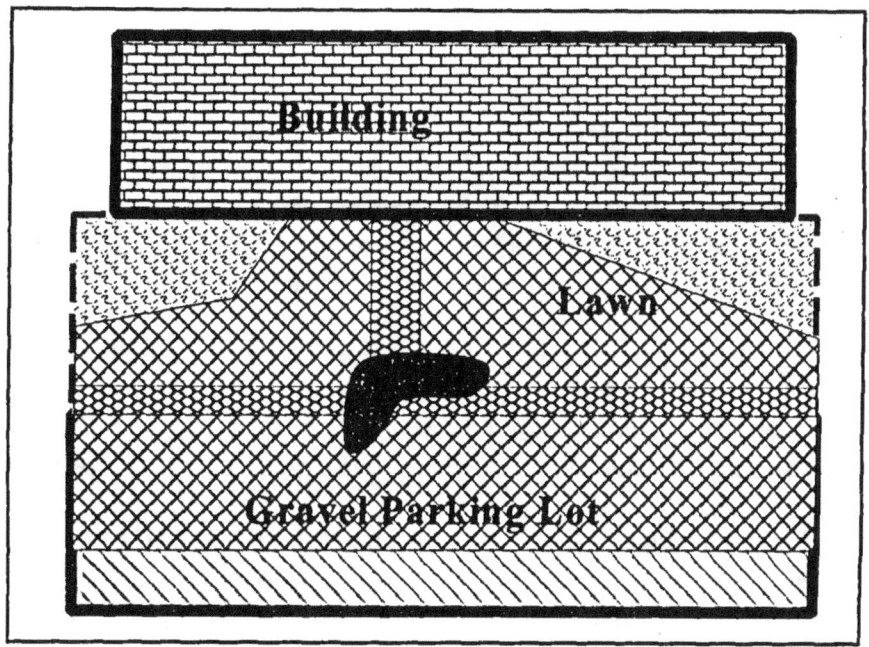

Figure 12.3 Example Survey Units: Case #3

These example cases are only meant to illustrate the considerations that may dictate how survey units may be designated, and what tradeoffs may be involved. What is actually done in any specific situation will depend on site-specific information from historical site assessments, and prior scoping, characterization, and remediation control survey results.

12.3 Using Paired Observations for Survey Units with Many Different Backgrounds

In this report, we have discussed using the Sign test for residual radioactivity that does not appear in background when radionuclide specific measurements are made, and otherwise using the WRS test. However, there are cases when one may wish to use the Sign test even when the radionuclide appears in background and/or radionuclide specific measurements are not made. An obvious instance would be when background is such a small fraction of the DCGL that including it is unlikely to affect the decision errors. An example would be ^{137}Cs residual radioactivity in an area where the concentrations from global fallout are small. It may be more cost-effective to simply compare the total ^{137}Cs concentration in the survey unit to the DCGL using the Sign test rather than to attempt to find a matching reference area.

Another case in which the Sign test may be more appropriate is when there are many different materials within what would otherwise logically be a single survey unit. As indicated at the beginning of this chapter, to divide such a survey unit into separate parts, each requiring its own reference area is not only impractical, but may be inconsistent with the dose models used to determine the DCGLs.

Consider once again, the example case #1 shown in Figure 12.1. Suppose the residual radioactivity of concern is ^{226}Ra rather than ^{60}Co. We will call this example case #4. When ^{60}Co was the concern, only the variability in a material's potential for retaining or accumulating this radionuclide was important. If ^{226}Ra is the concern, then the variability in the background concentration of ^{226}Ra in the materials is an additional, perhaps more important concern in forming survey units. In Figure 12.4, the same area as for case #1 is again shown, but with consideration of this additional factor.

With an eye towards potential differences in background ^{226}Ra, it becomes important that:
• The walkway was paved with different concrete at two different times.
• The parking lot was expanded using a different type of gravel.
• Part of the lawn was graded with fill from another location.
• Soil and mulch were used for the plant beds next to the building.

What were previously three potentially different survey units are now possibly as many as nine different survey units. The contamination potential is still Class 2, as in case #1, and on that basis alone this area might be designated as a single survey unit. This disparity of effort between case #1, using the Sign test for ^{60}Co, and case #4 using the WRS test for ^{226}Ra is tremendous.

Fortunately, there is a third option— to use the Sign test with paired observations. Each measurement in the survey unit is paired with an observation on a suitable reference material. The Sign test is then performed on the difference. The tradeoff is the higher variability of the differences compared to a single measurement.

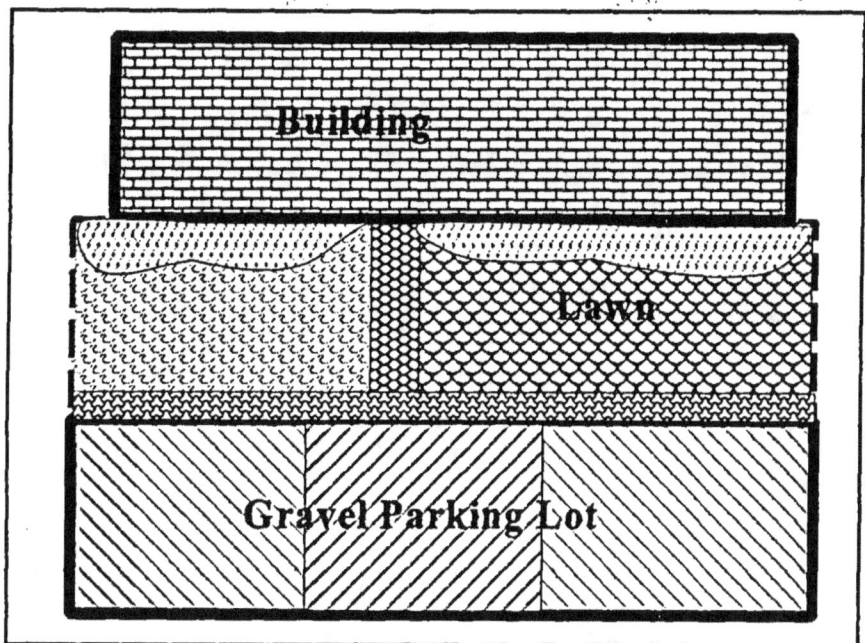

Figure 12.4 Example Survey Units: Case #4

In case #1, an estimate of the variability of the measurements from the survey unit, σ, is needed to determine the relative width of the gray area, Δ/σ, which then in turn is used to determine the required sample size, N. Suppose the survey unit measurements are designated by Y_i, for $i = 1$ to N. From each of these Y_i, a paired measurement, X_i, on an appropriate reference material is subtracted. The Sign test is performed on the difference $Y_i - X_i$. Thus it is the variability of these differences, $\sigma_{(Y_i - X_i)}$, that is required in order to determine the required sample size. This variability has three components:

$$\sigma^2_{Y_i - X_i} = \sigma^2_{Y_i} + \sigma^2_{X_i} = \sigma^2_R + \sigma^2_{B_i} + \sigma^2_{X_i} \tag{12-1}$$

where

σ_{X_i} is the standard deviation of the measurements on the reference material,
σ_{Y_i} is the standard deviation of the measurements in the survey unit,
σ_{B_i} is the standard deviation of the background in the survey unit material, and
σ_R is the standard deviation of the residual radioactivity in the survey unit.

If the reference material is truly representative, then $\sigma_{X_i} = \sigma_{B_i}$, so that

$$\sigma_{Y_i - X_i} = \sqrt{\sigma^2_R + 2\sigma^2_{B_i}} \tag{12-2}$$

Better precision may be possible if the average of m_j measurements made on the jth reference material is subtracted from each measurement from the survey unit made on that material.

The Sign test performed the differences:

$$Y_{j,k} - \overline{X}_j = \frac{Y_{j,k}}{m_j} \sum_{q=1}^{m_j} X_j \qquad (12\text{-}3)$$

for $k = 1$ to n_j.

The variability of these differences is

$$\sigma_{Y_{i,j} - \overline{X}_j} = \sqrt{\sigma_R^2 + \sigma_{B_j}^2 + \sigma_{B_j}^2/n_j} = \sqrt{\sigma_R^2 + \left(\frac{n_j+1}{n_j}\right)\sigma_{B_j}^2} \qquad (12\text{-}4)$$

Note that when ^{60}Co was the contaminant, the only component of variability was $\sigma = \sigma_R$.

To estimate the sample size needed for case #1, suppose the DCGL for ^{60}Co is 2 and that $\sigma_R = 0.7$. If the LBGR is set at 1, then $\Delta/\sigma = (2-1)/0.7 = 1.4$. For acceptable error rates of $\alpha = \beta = 0.05$, the sample size found in Table 3.2 is 20. Suppose, for the sake of illustration, that for case #4, the DCGL for ^{226}Ra is also 2, and $\sigma_R = 0.7$. If the ^{226}Ra background standard deviation is about 0.5, then standard deviation of the difference of matched pairs of measurements,

$$\sigma_{Y_i - X_i} = \sqrt{\sigma_R^2 + 2\sigma_{B_i}^2} = \sqrt{(0.7)^2 + 2(0.5)^2} = \sqrt{0.49 + 0.5} = \sqrt{0.99} \approx 1 \qquad (12\text{-}5)$$

Thus, $\Delta/\sigma = 1$, and for the same LBGR of 1 and acceptable error rates of $\alpha = \beta = 0.05$, the sample size found in Table 3.2 is 29 measurements in the survey unit. An additional matching 29 measurements on reference materials are also needed, for a total of 58 measurements. To simplify the above calculations, a single estimate of the standard deviation of background measurements was used for all materials. It would be prudent to use the largest anticipated standard deviation. Note that no assumption about the average of the background concentrations in the different materials was made. This may vary considerably from one material to another.

Indeed, if the average concentration does not vary significantly, it would be better to perform the WRS test using a reference area with a composition that is reasonably well-matched to the survey unit. Suppose in case #4, that the average concentrations of ^{226}Ra did not vary much. Then the standard deviation of background in the reference area is still about 0.5, and the variability in the survey unit is simply

$$\sigma_{Y_i} = \sqrt{\sigma_R^2 + \sigma_{B_i}^2} = \sqrt{(0.7)^2 + (0.5)^2} = \sqrt{0.49 + 0.25} = \sqrt{0.74} \approx 0.86 \qquad (12\text{-}6)$$

Thus, $\Delta/\sigma = 1.155$, and for the same LBGR of 1 and acceptable error rates of $\alpha = \beta = 0.05$, the sample size found by interpolating in Table 3.3 is about 26 measurements each required in the survey unit and reference area for a total of 52 measurements. The WRS test requires fewer measurements. An equivalent observation is that for the same number of measurements, the WRS test has greater power. The essential difference is whether the reference measurements can be considered independent of the survey unit measurements, or whether they must be matched together according to material type.

Notice that Equations 12-2, 12-4, and 12-6 differ primarily in the factor multiplying σ_B^2. Using the Sign test with a single matched reference measurement, this factor is 2. Using the WRS test, this factor is 1. If the mean of n_j measurements on the jth material is used this factor is $(n_j + 1)/n_j$. This equals 2 when $n_j = 1$, and approaches 1 as n_j becomes large.

The question remains as to how the measurements should be taken in the survey unit and from the reference materials. The measurements in the survey unit should be taken according to the regular procedure recommended for that class of survey unit, i.e., on a random start systematic grid for Class 1 and Class 2, and randomly for Class 3. This is essentially the same as sampling according to the proportional area of each material in the survey unit. Matching reference area samples should be taken randomly on the chosen reference material.

13 DEMONSTRATING INDISTINGUISHABILITY FROM BACKGROUND

Thus far in this report the emphasis has been on conducting final status surveys that demonstrate that any residual radioactivity in a survey unit is within the release criterion. In these cases, Scenario A is generally preferred for the survey design. In some cases, however, it may be more appropriate to demonstrate indistinguishability from background. Demonstrating indistinguishability from background using Scenario B will be a useful option when the residual radioactivity consists of radionuclides that appear in background, and the variability of the background is relatively high. Background variability may be considered high when differences in estimated mean concentration measured in potential reference areas are comparable to screening level DCGLs.

13.1 Determining Significant Background Variability

In Section 2.2.7, the concept of a reference area was introduced. Any difference in the concentrations between the reference area and the survey unit is assumed to be due to residual radioactivity. It is not possible to determine whether or not an observed difference is actually due to variations in the mean background concentrations between these areas.

When the variations in mean background among different potential reference areas are small compared to the width of the gray region, they can often be neglected. In such cases, the choice of reference area will not materially affect the decision on whether or not to release a survey unit to which it is compared.

As the variations in mean background among different potential reference areas become comparable in magnitude to the width of the gray region, they can no longer be ignored. When the reference area has a higher mean background than the survey unit, the survey unit will be more likely to pass, and when the reference area has a lower mean background than the survey unit, the survey unit will be more likely to fail. Since any difference in background activity between the survey unit and the reference area is attributed to residual radioactivity, the choice of reference area may materially affect the decision on whether or not to release a survey unit to which it is compared.

As an example, consider Figure 13.1, which illustrates a DQO specification for a survey design. The gray region and acceptable rates for decision errors are shown by the solid curve. Suppose the reference area happens to have a lower mean concentration than the actual background concentrations in the survey unit. This difference is depicted by the double-headed arrow. The values of residual radioactivity concentrations in this survey unit will appear larger than they actually are by the amount of that difference. The result is that the actual probability that the survey unit passes is represented by the dashed curve, which is shifted downward from the solid curve by the difference in mean background between the survey unit and the reference area. This means, for example, that when the true residual radioactivity concentration is at the LBGR there is only about a 65% probability of passing this survey unit, rather the 95% probability specified in the DQO.

Exactly how much, and in which direction the probabilities shift will depend on the particular reference area. Under such circumstances, whether a survey unit passes or fails may depend more on the particular reference area chosen than on the amount of residual radioactivity that it contains. This leads to a quantitative definition of what it means for a survey unit to be indistinguishable from background. It is expressed in terms of the potential for variations in reference area mean concentrations to impact decision error rates. First, it is necessary to establish that there is significant variability among potential reference areas. A procedure for doing this is discussed in the next section. If it shown that significant variability exists, this information is used to define a level of residual radioactivity concentration that is indistinguishable from background variations. Section 13.4 discusses how this can be used to plan a final status survey using Scenario B.

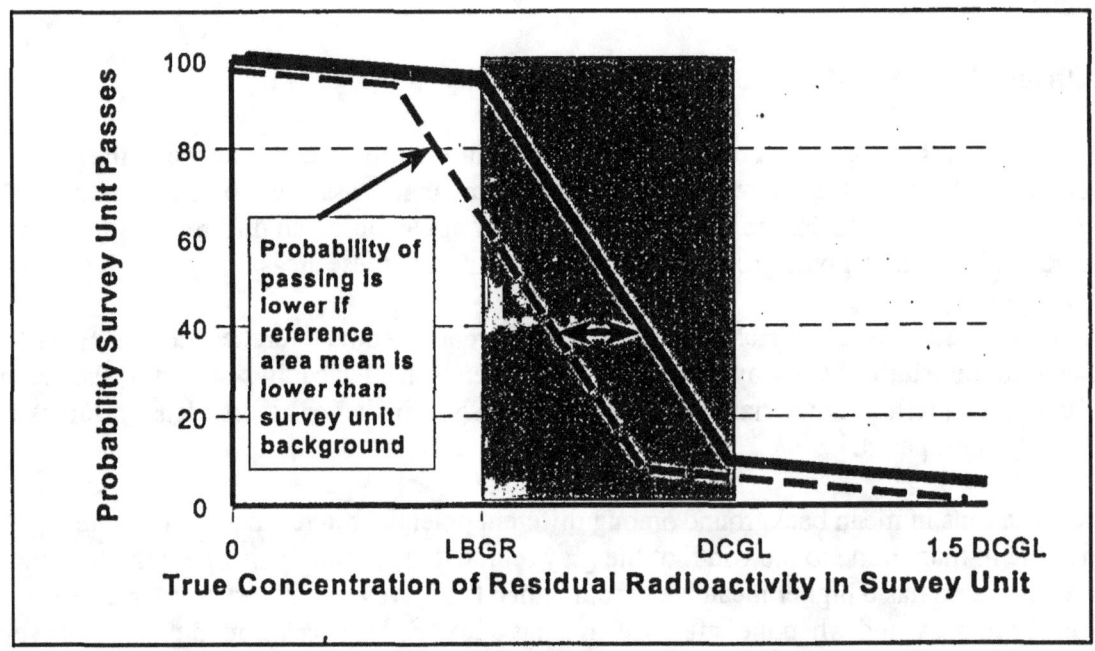

Figure 13.1 Impact of Background Variability on Decision Errors

13.2 Determining if Reference Areas Have Significantly Different Background Levels

In this section, we focus our attention on potential reference areas. Consider all the reference areas to which a particular survey unit may be compared according to the criteria set out in Section 2.2.7. To determine whether there are significant differences among the reference area background means, the reference area measurements are expressed as:

$$x_{ij} = \zeta + \mu_i + z_{ij} \tag{13-1}$$

where

x_{ij} = the jth measurement in the ith reference area, for $j = 1$ to n_i, the number of measurements in the ith reference area, and for $i = 1$ to k, the number of reference areas

ζ = the mean concentration over all reference areas

μ_i = the difference between the overall mean and the mean in the ith reference area

z_{ij} = the contribution of random spatial and measurement variability to the jth measurement in the ith reference area

ξ is an unknown constant. The μ_i are distributed across reference areas with mean zero and standard deviation ω, but within reference area i, μ_i has a fixed value. The z_{ij} have mean zero and standard deviation σ. The measurements within reference area i have mean $\xi + \mu_i$, and variance σ^2. The reference area means are distributed around the overall mean ξ with variance deviation ω^2. Thus, if there is no variability in the reference area means, then $\omega^2 = 0$. The measurement variability within each reference area is σ^2, and it is the same whether or not there is a significant difference among the reference area means.

If the μ_i and the z_{ij} are assumed to be normally distributed, the above corresponds to a random effects one-way analysis of variance model, sometimes called Model II. ω^2 and σ^2 are called the components of variance. The null hypothesis, $H_0 : \omega^2 = 0$, versus the alternative, $H_a : \omega^2 > 0$, is tested parametrically with an F-test. The non-parametric equivalent is the Kruskal-Wallis test.

Before collecting data for the Kruskal-Wallis test, the acceptable Type I error rate, α_{KW}, must be specified. This is the acceptable probability of concluding that the reference areas have different average concentrations, when in fact they are the same. In setting α_{KW} it is important to consider that the risk involved in a Type I error may be much smaller than the risk of a Type II error.

To perform the Kruskal-Wallis test, all of the measurements from the reference areas are pooled and ranked. For every measurement, x_{ij}, there is a corresponding rank, r_{ij}. There will be $N = n_1 + n_2 + ... + n_k$ measurements in all. The sum of all of the ranks is $N(N+1)/2$. Therefore the average rank is $N(N+1)/(2N) = (N+1)/2$. If the distribution of measured concentrations in each reference area is the same, then the average rank for each reference area, should also be about the same, i.e., $(N+1)/2$.

Let

$$\overline{R}_i = \frac{1}{n_i} \sum_{j=1}^{n_i} r_{ij} = R_i/n_i \qquad (13\text{-}2)$$

be the average rank in reference area i. The quantity $\overline{R}_i - (N+1)/2$ is a measure of how different this reference area is from the others.

The Kruskal-Wallis statistic is a weighted sum of the squares of these differences over all of the reference areas:

$$K = \frac{12}{N(N+1)} \sum_{i=1}^{k} n_i \left(\overline{R}_i - (N+1)/2 \right)^2 = \frac{12}{N(N+1)} \left(\sum_{i=1}^{k} R_i^2/n_i \right) - 3(N+1) \qquad (13\text{-}3)$$

The weights in Equation 13-3 have been chosen so that the probability that K exceeds a given value K_c, may be approximated by a chi-squared distribution with $k-1$ degrees of freedom:

$$\text{Prob}(K > K_c) = 1 - \chi^2_{k-1}(K_c) \qquad (13\text{-}4)$$

NUREG-1505

The critical value K_c is determined from setting $\chi^2_{k-1}(K_c) = \alpha_{KW}$. Values of $\chi^2_{k-1}(K_c)$ for typical values of α_{KW} are tabulated in Table 13.1. If the value of K computed from Equation 13-3 exceeds K_c, then the null hypothesis is rejected.

Table 13.1 Critical Values, K_c, for the Kruskal-Wallis Test

$k-1$	α_{KW}				
	0.01	0.025	0.05	0.1	0.2
1	6.6	5.0	3.8	2.7	1.6
2	9.2	7.4	6.0	4.6	3.2
3	11.3	9.3	7.8	6.3	4.6
4	13.3	11.1	9.5	7.8	6.0
5	15.1	12.8	11.1	9.2	7.3
6	16.8	14.4	12.6	10.6	8.6
7	18.5	16.0	14.1	12.0	9.8
8	20.1	17.5	15.5	13.4	11.0
9	21.7	19.0	16.9	14.7	12.2
10	23.2	20.5	18.3	16.0	13.4

For example, suppose there are four reference areas under consideration, and ten measurements are made in each. The data and the ranks are shown in Table 13.2. The same type of spreadsheet functions that were used for the WRS test can also be used to calculate the ranks for the Kruskal-Wallis test.

Using Equation 13-2,

$$K = \frac{12}{N(N+1)} \left(\sum_{i=1}^{k} R_i^2/n_i \right) - 3(N+1)$$

$$= \frac{12}{40(41)} \left(155^2/10 + 121^2/10 + 255^2/10 + 289^2/10 \right) - 3(41)$$

$$= \frac{12}{1640} (2402.5 + 1464.1 + 6502.5 + 8352.1) - 123 = \frac{12}{1640} (18721.2) - 123.0 = 14.0$$

With $k-1 = 3$, this value of K, is greater than the highest value of K_c in Table 13.1, viz., 11.3 for $\alpha_{KW} = 0.01$. The null hypothesis is rejected. It is concluded that these reference areas do have significantly different concentration distributions.

Table 13.2 Example Data for the Kruskal-Wallis Test

	Measurements					Ranks			
	Area 1	Area 2	Area 3	Area 4		Area 1	Area 2	Area 3	Area 4
1	0.27	1.04	2.45	3.77		6	13	27	39
2	1.87	0.39	0.34	2.63		20	9	8	31
3	0.97	2.07	3.06	4.05		10	23	37	40
4	1.01	−0.57	2.83	1.72		11	2	35	19
5	2.08	1.97	1.09	1.50		24	21	14	17
6	1.62	−0.22	0.26	2.47		18	3	5	29
7	0.30	1.39	2.80	1.42		7	15	34	16
8	1.98	0.05	2.77	2.47		22	4	33	28
9	2.18	−0.75	2.42	2.76		25	1	26	32
10	1.02	2.50	2.86	3.35		12	30	36	38
Mean	1.33	0.79	2.09	2.61	Sum	155	121	255	289
StdDev	0.71	1.17	1.09	0.91	Total				820

13.3 Establishing the Concentration Level That Is Indistinguishable

Once it is decided that there are significant differences among the potential reference areas for a survey unit, a measure of the variability among these reference areas is needed.

The sample mean of the measurements in area i is

$$\overline{x}_i = \frac{1}{n_i} \sum_{j=1}^{n_i} x_{ij} \tag{13-5}$$

which provides an estimate of the mean concentration in reference area i, namely $\zeta + \mu_i$.

The overall sample mean

$$\overline{x} = \frac{\sum_{i=1}^{k} \sum_{j=1}^{n_i} x_{ij}}{\sum_{i=1}^{k} n_i} \tag{13-6}$$

NUREG-1505

is an estimate of the overall mean background concentration, ξ.

The sample variance of the measurements in area i is

$$s_i^2 = \sum_{j=1}^{n_i} (x_{ij} - \bar{x}_i)^2 / (n_i - 1) = \sum_{j=1}^{n_i} (z_{ij} - \bar{z}_i)^2 / (n_i - 1) \tag{13-7}$$

which is an estimate of σ^2.

Since σ^2, is assumed to be the same in each reference area, these estimates can be pooled into the following estimate:

$$s_w^2 = \frac{\sum_{i=1}^{k} \sum_{j=1}^{n_i} (x_{ij} - \bar{x}_i)^2}{\sum_{i=1}^{k} (n_i - 1)} = \frac{\sum_{i=1}^{k} \sum_{j=1}^{n_i} x_{ij}^2 - \sum_{i=1}^{k} n_i (\bar{x}_i)^2}{\sum_{i=1}^{k} (n_i - 1)} \tag{13-8}$$

In the analysis of variance, s_w^2 is called the mean square within reference areas [1].

The mean square between reference areas is

$$s_b^2 = \frac{\sum_{i=1}^{k} n_i (\bar{x}_i - \bar{x})^2}{k - 1} = \frac{\sum_{i=1}^{k} n_i (\bar{x}_i)^2 - \left(\sum_{i=1}^{k} \sum_{j=1}^{n_i} x_{ij} \right)^2 / \sum_{i=1}^{k} n_i}{k - 1}$$

The righthand sides of Equations 13-8 and 13-9 may appear imposing, but essentially only involve:

(1) the sum of the squares of all the measurements,

$$\sum_{i=1}^{k} \sum_{j=1}^{n_i} x_{ij}^2 \tag{13-10}$$

---

[1] The sum of the squared differences from the overall mean is composed of two parts:

$$\sum_{i=1}^{k} \sum_{j=1}^{n_i} (x_{ij} - \xi)^2 = \sum_{i=1}^{k} \sum_{j=1}^{n_i} (x_{ij} - \bar{x}_i)^2 + \sum_{i=1}^{k} \sum_{j=1}^{n_i} (\bar{x}_i - \xi)^2 = \sum_{i=1}^{k} \sum_{j=1}^{n_i} (x_{ij} - \bar{x}_i)^2 + \sum_{i=1}^{k} n_i (\bar{x}_i - \xi)^2$$

The first term on the right hand side is the sum of squares within reference areas and the second term on the right hand side is the sum of squares between reference areas. The mean square is obtained by dividing the sum of squares by the degrees of freedom. For the means square within reference areas this is the total number of data points minus the number of reference areas. For the means square between reference areas this is the number of reference areas minus one.

(2) the square of the sum of all the measurements,

$$\left(\sum_{i=1}^{k}\sum_{j=1}^{n_i} x_{ij}\right)^2 \qquad (13\text{-}11)$$

and

(3) the sum of the squares of the reference area averages weighted by the number of measurements,

$$\sum_{i=1}^{k} n_i \left(\bar{x}_i\right)^2 \qquad (13\text{-}12)$$

The component of variance, ω^2, is estimated by

$$\omega^2 = (s_b^2 - s_w^2)/n_0 \quad \text{where} \quad n_0 = \frac{N - \sum_{i=1}^{k} n_i^2 / N}{(k-1)} \qquad (13\text{-}13)$$

n_0 is usually slightly less than the average number of samples taken in each reference area,

$$\bar{n} = \frac{1}{k}\sum_{i=1}^{k} n_i \qquad (13\text{-}14)$$

If the number of measurements in each reference area is the same, $n_1 = n_2 = ... = n_k = n$, then,

$$n_0 = \frac{N - \sum_{i=1}^{k} n_i^2 / N}{(k-1)} = \frac{N - kn^2/N}{(k-1)} = \frac{kn - kn^2/kn}{(k-1)} = \frac{kn - n}{(k-1)} = \frac{n(k-1)}{(k-1)} = n \qquad (13\text{-}15)$$

The calculation of ω^2 for the example data of Table 13.2 proceeds using the sums and squares shown in Table 13.3.

The sums (1), (2), and (3) together with s_b^2 and s_w^2 are calculated from Table 13.3 as follows:

From Equation 13-10,
(1) = sum of squares = $22.28 + 18.50 + 54.30 + 75.80 = 170.88$.

From Equation 13-11,
(2) = square of the sum = $(13.30 + 7.87 + 20.88 + 26.14)^2 = (68.19)^2 = 4649.88$.

From Equation 13-12,

(3) = weighted sum of the squares of the averages $10(1.33^2 + 0.79^2 + 2.09^2 + 2.61^2)$

$= 10(1.77 + 0.62 + 4.36 + 6.83) = 10(13.58) = 135.8.$

From Equation 13-8,

$s_w^2 = (170.88 - 135.8)/(N - k) = (170.88 - 135.8)/36 = 0.97.$

From Equation 13-9,

$s_b^2 = (135.8 - (4649.88/N))/(k - 1) = (135.8 - (4649.88/40))/(3) = 6.52.$

Finally, we have

$$\hat{\omega}^2 = (s_b^2 - s_w^2)/n_0 = (6.52 - 0.97)/10 = 0.55 \tag{13-16}$$

Table 13.3 Calculation of $\hat{\omega}^2$ for the Example Data

	Measurements				Measurements Squared			
	Area 1	Area 2	Area 3	Area 4	Area 1	Area 2	Area 3	Area 4
1	0.27	1.04	2.45	3.77	0.07	1.08	6.00	14.21
2	1.87	0.39	0.34	2.63	3.50	0.15	0.12	6.92
3	0.97	2.07	3.06	4.05	0.94	4.28	9.36	16.40
4	1.01	-0.57	2.83	1.72	1.02	0.32	8.01	2.96
5	2.08	1.97	1.09	1.50	4.33	3.88	1.19	2.25
6	1.62	-0.22	0.26	2.47	2.62	0.05	0.07	6.10
7	0.30	1.39	2.80	1.42	0.09	1.93	7.84	2.02
8	1.98	0.05	2.77	2.47	3.92	0.00	7.67	6.10
9	2.18	-0.75	2.42	2.76	4.75	0.56	5.86	7.62
10	1.02	2.50	2.86	3.35	1.04	6.25	8.18	11.22
sum	13.30	7.87	20.88	26.14	22.28	18.50	54.30	75.80
average	1.33	0.79	2.09	2.61				
avg sqd	1.77	0.62	4.36	6.83				

Although the analysis of variance using the F-test requires the assumption that the data are normally distributed, the calculation of $\hat{\omega}^2$ does not. Therefore, the values of the mean squares s_b^2 and s_w^2 that are generated by most statistical computer programs for ANOVA can be used for these calculations. Table 13.4 shows an ANOVA for the example data generated by a spreadsheet program. The entry for the mean square within groups, 0.97, is the same as was found in Table 13.3 for s_w^2. Similarly, the entry for the mean square between groups, 6.52, is the same as was found in Table 13.3 for s_b^2. The F-statistic, which is simply the ratio s_b^2/s_w^2, is also shown. Of course, if the data from each reference area are consistent with the assumption of normality, the F-test may simply be used instead of the Kruskal-Wallis test.

Table 13.4 Analysis of Variance for Example Data

Source of Variation	Sum of Squares	Degrees of Freedom	Mean Square	F Statistic
Between Groups	19.56	3	6.52	6.69
Within Groups	35.08	36	0.97	
Total	54.65	39		

13.4 Using the Concentration Level That Is Indistinguishable in the WRS Test

Recall from Section 3.6, that in Scenario B, the hypotheses being tested are

Null Hypothesis:
H_0: The mean concentration of residual radioactivity in the survey unit is indistinguishable from background up to a level specified by the LBGR.
versus
Alternative Hypothesis:
H_a: The mean concentration of residual radioactivity in the survey unit distinguishable from background is in excess of the $DCGL_W$.

In this scenario, a Type I decision error, with associated probability α, is made when a survey unit fails when it should pass. A Type II decision error, with associated probability β, is made when a survey unit passes when it should fail. To set these decision errors, an appropriate gray region is needed. The lower bound on this gray region is the concentration level above background that may be considered distinguishable from background.

If the null hypothesis of Kruskal-Wallis test has been rejected, the mean background levels among reference areas varies about the overall mean ξ with a standard deviation estimated by $\hat{\omega}$. The difference in concentration that is distinguishable above background variability may be expressed in terms of an appropriate multiple of $\hat{\omega}$. For example, if the reference area means are normally distributed, the probability that the survey unit mean is more than two standard deviations away from the overall mean is about 5%. Regardless of how the data are distributed, Chebyshev's Inequality states that the probability that the true mean background in the survey unit differs from the overall mean background by more than t standard deviations is less than $1/t^2$. Therefore, the probability that the survey unit mean is more than $t = 2$ standard deviations away from the overall mean is less than $1/t^2 = 1/4 = 25\%$. The probability that the survey unit mean is more than $t = 3$ standard deviations away from the overall mean is less than $1/t^2 = 1/9 = 11\%$. In most cases, it is reasonable to assume that the true probabilities will fall somewhere between the value calculated for the normal distribution and that established by Chebyshev's inequality. The multiple that of $\hat{\omega}$ that is used in a specific application should be decided during the DQO process, but a factor of three is a reasonable default.

For the example data $\hat{\omega}^2 = 0.55$ so $\hat{\omega} = 0.74$. Thus, differences smaller than $3\hat{\omega} = 2.22$ would not be considered distinguishable from background variations. Notice that in Table 13.2, the difference in the means between reference areas #4 and #2 is $2.61 - 0.79 = 1.82$.

The WRS test is applied as described in Section 6.3, using the decided upon multiple of $\hat{\omega}$ as the LBGR, and the width of the gray region equal to the $DCGL_w$. The hypotheses tested by the WRS under Scenario B are restated as

Null Hypothesis:
H_0: The difference in the median concentration of radioactivity in the survey unit and in the reference area is less than the LBGR.
versus
Alternative Hypothesis:
H_a: The difference in the median concentration of radioactivity in the survey unit and in the reference area is greater than the $DCGL_w$.

The Type I error rate $\alpha_w = \alpha/2^{(2)}$, is the probability that a survey unit with a difference from the reference area equal to the LBGR will fail the test. The power, $1 - \beta$, is the probability that a survey unit with a difference at the $DCGL_w$ above the LBGR will fail the test. For example, the desired probability for the survey unit passing might look similar to Figure 13.2.

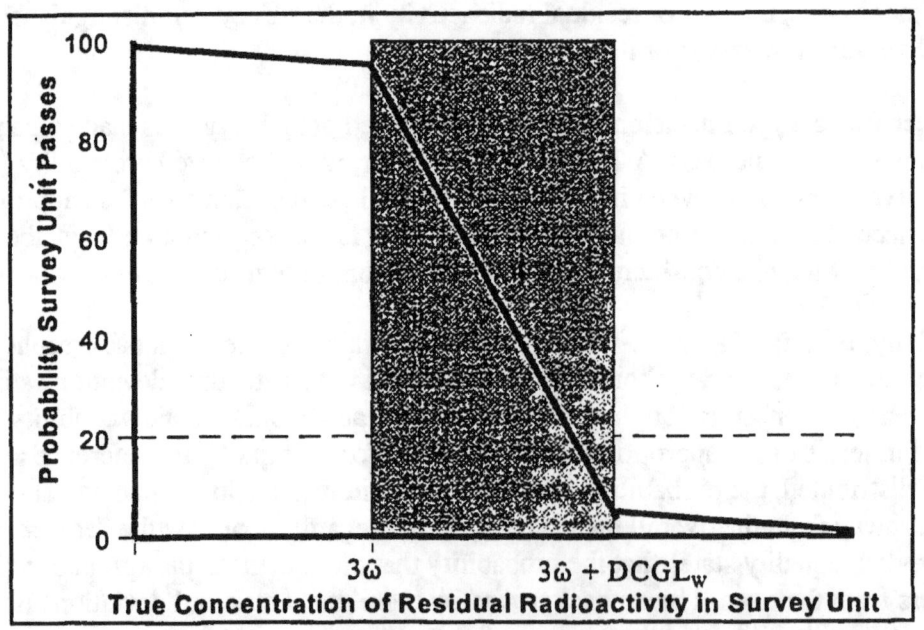

Figure 13.2 Example DQOs for the Probability That the Survey Unit Passes Versus the Concentration Difference Between the Survey Unit and the Reference Area

---

[2] Recall that since the Quantile test is performed in tandem with the WRS test, $\alpha_w = \alpha_Q = \alpha/2$, so that the that the size of the two tests in tandem is approximately $\alpha = \alpha_Q + \alpha_w$.

The Quantile test is also performed as described in Section 7.2, with $\alpha_Q = \alpha/2$, and with the LBGR equal to the decided upon multiple of $\hat{\omega}$.

All of the reference area measurements taken for the Kruskal-Wallis test should also be used in the WRS and Quantile tests. In most cases, no additional reference area samples will be needed. If additional samples are required, they should be obtained randomly, with all of the reference areas being equally likely to be sampled.

13.5 Determining the Number of Reference Areas and the Number of Samples

In applying the methods of this chapter, it is natural to ask is how many reference areas should be studied, and how many samples should be taken in each. As was seen in Section 3.8.1, the number of samples needed depends on both the probability of a Type I error (α) and the probability of a Type II error (β) that are deemed acceptable for the test. Unfortunately, the power ($1-\beta$) of the Kruskal-Wallis test involves functions that are "too complicated to be useful" (Lehmann and D'Abrerra, 1975). However, it has been shown that the efficiency of the Kruskal-Wallis test relative to the F-test is the same as the efficiency of the WRS test to the t-test (Andrews, 1954). This means that one can get an approximate idea of the power of the Kruskal-Wallis test by calculating the power of the F-test.
The power of the F-test is (Brownlee, 1960, p.268):

$$1-\beta = Probability[F(f_1,f_2) > \frac{1}{\phi} F_{1-\alpha}(f_1,f_2)] \tag{13-17}$$

where

$$\phi = 1 + n\frac{\omega^2}{\sigma^2} \text{ , and } F_{1-\alpha}(f_1,f_2) \text{ is the } 1-\alpha_{KW} \text{ percentile of the } F \text{ distribution with } f_1 = k - 1$$

and $f_2 = kn - k$ degrees of freedom. Through ϕ, the power depends on the ratio of the variance components. Under the null hypothesis, $\omega^2 = 0$, so $\phi = 1$.

As an example, consider the data from Table 13.2. If one wished to detect a situation in which $\omega^2 = \sigma^2$, then $\phi = 1 + 10(1) = 11$. If $\alpha_{KW} = 0.1$, then $F_{1-\alpha}(f_1,f_2) = F_{0.9}(3,36) = 2.243$. The

power, $1-\beta$, is then the probability that the F-statistic with 3 and 36 degrees of freedom exceeds $2.243/\phi = 2.243/11 = 0.2039$. β is the probability that the F-statistic with 3 and 36 degrees of freedom is less than 0.2039. This probability is about $\beta = 0.1$, so the power is about 0.9.

The results of this calculation for other numbers of samples and reference areas are shown in Table 13.5 for $\alpha_{KW} = 0.05, 0.10$ and 0.20. Although this is only an approximation, and the actual power of the Kruskal-Wallis test would be slightly lower, this table indicates that with four reference areas each with between 10 and 20 samples in each should generally be adequate. Since the risk of not detecting background variations that are actually present (a Type II error) could involve the impossible task of remediating background, choosing a higher value for α_{KW} than for β would often be justified. From Table 13.5, when $k = 4$, this implies that $\alpha_{KW} = 0.1$ is a reasonable default, and in some circumstances even larger values could be considered.

Table 13.5 Power of the F-test When $\omega^2 = \sigma^2$

Number of reference areas (k)	Number of samples in each (n)	Total Number of Samples (kn)	ϕ	Power when $\alpha_{KW} = 0.05$	Power when $\alpha_{KW} = 0.1$	Power when $\alpha_{KW} = 0.2$
2	10	20	11	53.4%	60.7%	69.3%
2	15	30	16	61.3%	67.4%	74.5%
2	20	40	21	66.1%	71.5%	77.7%
2	30	60	31	72.1%	76.5%	81.7%
3	10	30	11	74.0%	79.7%	85.7%
3	15	45	16	81.8%	85.9%	90.1%
3	20	60	21	86.1%	89.2%	92.4%
3	30	90	31	90.5%	92.7%	94.9%
4	10	40	11	85.3%	89.3%	93.0%
4	15	60	16	91.4%	93.8%	96.0%
4	20	80	21	94.2%	95.8%	97.3%
4	30	120	31	96.7%	97.6%	98.5%
5	10	50	11	91.8%	94.3%	96.5%
5	15	75	16	95.9%	97.2%	98.3%
5	20	100	21	97.6%	98.4%	99.0%
5	30	150	31	98.9%	99.2%	99.5%
6	10	60	11	95.4%	97.0%	98.3%
6	15	90	16	98.1%	98.8%	99.3%
6	20	120	21	99.0%	99.4%	99.6%
6	30	180	31	99.6%	99.8%	99.9%

13.6 Determining When Demonstrating Indistinguishability Is Appropriate

The methods of this chapter were developed specifically to address potential difficulties with demonstrating compliance with dose-based release criteria at sites with spatially variable background concentrations of natural radionuclides. Generally, the use of Scenario A is preferred since it involves fewer assumptions in its application and requires only one statistical test. The null hypothesis for Scenario A is such that when the residual radioactivity in the survey unit is very close to the $DCGL_W$, the default decision is to not release the survey unit without further investigation. This provides additional assurance that the release criteria will not be exceeded. However, when the variability in background is high, this assurance comes at too high a price, namely the possibility of requiring remediation of survey units containing only background concentrations of radionuclides. Since the default decision in Scenario B is to release the survey unit, a second statistical test, the Quantile test, is used to detect non-uniform concentrations of residual radioactivity that may be excess of the release criterion, but that might be missed by the WRS test..

It is not possible to anticipate every circumstance in which these methods might be considered applicable. The suitability of these methods to specific situations should be determined during the DQO process. Two factors that should be considered in making this determination are:

(1) Have reasonable efforts been made to reduce measurement uncertainty. e.g., by use of radionuclide-specific methods?

(2) Have reasonable efforts been made to reduce spatial variability by choosing homogeneous survey units with well-matched reference areas?

Once it is determined that the methods of this chapter are appropriate, the error rates for the Kruskal-Wallis test should be set. The Kruskal-Wallis test is used to determine whether the spatial variability of average reference area background concentrations is significant. The significance level is . If adequate consideration has been given to the decision to demonstrate indistinguishability, α_{KW} need not be set to too low a value. Indeed, if it is felt that background variability should always have the benefit of the doubt, the Kruskal-Wallis test need not be conducted. Not conducting the test is essentially the same as setting $\alpha_{KW} = 1.0$. Table 13.5 could still be used as a guide in determining the number of reference areas, and the number of measurements in each, that will be used to estimate $\hat{\omega}$ according to the procedures of Section 13.3.

Finally, the appropriate multiple of $\hat{\omega}$ to be used as the LBGR should be determined. The discussion of this issue in Section 13.4 can be used as a guide.

14 ALTERNATIVES AND MODIFICATIONS

14.1 Alternative Statistical Tests

The nonparametric statistical tests described in this report are expected to perform well in a wide variety of circumstances. However, in some situations alternative methods can be considered. As mentioned in Section 2.4.1, there are many statistical tests that can be used for determining whether or not a survey unit meets the release criteria. Any one test may perform better or worse than others, depending on the hypotheses to be tested, i.e., the decision that is to be made and the alternative, and how well the assumptions of the test fit the situation. Some possible alternatives are discussed below.

In evaluating statistical tests, generally one chooses the test that has the highest power among the various alternatives. The power is compared when each test is set to have the same Type I error rate, α. The Type I error rate is the probability that the null hypothesis will be rejected when it is true. If the assumptions made about the data distribution are correct, the calculation of α forms the basis for setting the critical value of the test statistic. If the assumptions are not valid, the calculated value of α will differ from the true Type I error rate. The fewer assumptions that are made, the more confidence can be placed in the calculation of the Type I error rate.

If a specific set of assumptions is made, the test results can be simulated using Monte Carlo sampling techniques. Using a large number of simulations, the actual Type I and Type II error rates for different tests can be compared. For each sample size, and specific set of assumptions, a separate simulation must be performed. Although much can be learned about the relative accuracy of statistical tests in this way, it is clearly not possible to explore every potential set of assumptions.

An alternative is to look at large sample results. With very few exceptions, it can be proved that the average of a large enough number of random data points tends to be normally distributed (Central Limit Theorem). In the same way, the power of statistical tests can be examined when the sample sizes are large enough. Note that what is meant by large enough is not precisely specified. Depending on the situation, large enough might be 10, or it might be 1000. If the samples size is allowed to grow large enough, the asymptotic (i.e., in the limit of arbitrarily large sample size) behavior of tests can be compared. Better large sample test behavior may be taken to imply that a test is better for all sample sizes. In reality, it can only be used as an indication of which test might be preferred.

One measure commonly used to compare statistical tests is called the relative efficiency. This is defined as the inverse of the ratio of sample sizes needed to achieve a given level of statistical power. If a test has relative efficiency of two relative to another, it requires half the sample size to achieve the same power. The asymptotic relative efficiency of one test to another, is the limit of the relative efficiency when the sample size is arbitrarily large.

Wilcoxon Signed Ranks Test (WSR test)

The asymptotic relative efficiency of the WSR test compared to the Sign test can be greater or less than one. That is, either might be better, depending on the data distribution. The WSR test tends to be better when the data distribution is symmetric, and the Sign test tends to be better when the data distribution is skewed.

Student's t-Test

Student's t-test may be used if the data have a normal distribution. This is a more restrictive requirement than that of symmetry, since every normal distribution is symmetric, but there are many other distributions that are also symmetric. The assumption of normality should be checked before using this test. The Shapiro-Wilk test discussed in EPA/QA-G9 (1996) is one such test. Others include the Kolmogorov-Smirnov test, Lillifor's test, and the Chi-Squared test.

The asymptotic relative efficiency of the WSR test relative to the one-sample Student's t-test ranges from 0.864 to infinity. As stated by Conover (1980): "the Wilcoxon test never can be too bad, but it can be infinitely good... ." The asymptotic relative efficiency of the WRS test compared to the two-sample Student's t-test has the same range, from 0.864 to infinity.

Chen's Test

Chen's test (Chen, 1995) is a modification of the Student's t-test that has been suggested for use when data are from a positively skewed distribution. Simulations show that it is generally more powerful than other forms of the t-test. However, this test can only be used in Scenario B.

Lognormal Test

If the data are assumed to lognormal, the testing procedure of Land (1988) may be used. The assumption of lognormality should be checked by testing the logarithms of the data for normality. It is important to note that a test on the mean of a lognormal distribution *cannot* be performed by using a Student's t-test on the mean of the logarithms of the data. This is because the mean of the logarithms of the data is the logarithm of the *median* of the original data. The behavior of this test relative to others when the assumption of lognormality is violated has not been studied.

Bootstrap Methods

The bootstrap is a simulation technique (Efron and Tibshirani, 1993). In essence, the distribution of concentrations in a survey unit is approximated by the empirical distribution (e.g., histogram) of the sample data taken. If n measurements are made, these n measurements are randomly sampled n times with replacement. Each time this is done, the mean of the random sample is calculated. After this has been done a specified number of times (generally between 50 and 200), the standard deviation of all of the random sample means is calculated. This is then used as the estimate of the standard error of the mean. There are, in addition, several methods for computing bootstrap t-statistics. Usually 1000 or more replications are recommended for the bootstrap t. Bootstrap methods generally have good asymptotic properties, but can be sensitive to outliers, and erratic when sample sizes are small.

14.2 Retesting

It may happen that a survey unit fails the hypothesis test (i.e., the decision is made that the survey unit does not meet the release criterion), yet the mean of the measured data is below the release criterion. This is more likely to occur when the mean falls in the gray region than otherwise. It analogous to the situation in which the mean is below the release criterion, but the $1-\alpha$ upper confidence level on the mean falls above the release criterion. It may be that the survey unit does meet the release criterion, but the hypothesis test was not powerful enough to detect that with the number of samples taken. Under some circumstances, one might like the option to take additional random samples and re-perform the hypothesis test on the entire set of data. The major difficulty with this is that the Type I error rate will now be greater than originally specified in the DQOs.

Sequential testing is performed when data are collected and analyzed in stages. It differs from hypothesis testing in that at each stage a third alternative is added to the decision of whether or not to reject the null hypothesis, namely, to collect more data before deciding. The usual motivation for sequential testing is to reduce the expected total number of samples from that required when all the sample are taken at one stage. Sequential versions of the WSR and WRS tests are discussed by Spurrier and Hewett (1976).

14.3 Composite Sampling

The number of measurements taken in a Class 1 survey unit may sometimes be driven more by the need to locate small areas of elevated activity than by the need to achieve the specified acceptable error rates for the statistical tests. When the scanning MDC is high, the sample size, N, may need to be significantly increased, in order to decrease the area between samples on the systematic grid. When this grid area, about A/N, is small enough, so that in turn the area factor is sufficiently high, the result is a $DCGL_{EMC}$ that is detectable by scanning. If the sample size N is much greater than that required for the statistical tests, some number, m, of neighboring samples might be composited to reduce the total cost of analysis. Suppose there are $N = mn$ measurements. Each composite represents a contiguous area of approximately the same proportion of the survey unit, $m(A/N)$. The number of composite measurements, n, should be equal to or greater than the number of measurements required by the statistical test. When the elevated measurement comparison is performed against the composites measurement results, the $DCGL_{EMC}$ should be divided by the number of samples included in each composite. If the composite measurement is below $DCGL_{EMC}/m$, no individual sample contributing to the composite could exceed the $DCGL_{EMC}$.

If a composite measurement is flagged by the EMC, it may be necessary to reanalyze each sample included in that composite to determine which of them, if any, actually exceed the $DCGL_{EMC}$ or, alternatively, the area of the survey unit represented by that composite measurement should be reinvestigated.

15 GLOSSARY

A_{min}: The smallest *area of elevated activity* identified using the DQO Process that it is important to identify.

action level: A scanning measurement level for residual radioactivity that (1) is based on the DCGL and (2) triggers a response, such as further investigation or cleanup, if exceeded. See *investigation level*.

activity: See *radioactivity*.

affected area: This term was previously used for areas that would be designated *Class 1 or Class 2*.

affected/non-uniform: This term was previously used for areas that would be designated *Class 1*.

affected/uniform: This term was previously used for areas that would be designated *Class 2*.

ALARA (acronym for as low as is reasonably achievable): A basic concept of radiation protection which specifies that exposure to ionizing radiation and releases of radioactive materials should be reduced as far below regulatory limits as is reasonably achievable considering economic, technological, and societal factors, among others. Reducing exposure at a site to *ALARA* strikes a balance between what is possible through additional remediation and the use of additional resources to achieve a lower level. A determination of *ALARA* is a site-specific analysis that is open to interpretation, because it depends on approaches or circumstances that may differ between regulatory agencies. An *ALARA* recommendation should not be interpreted as a set limit or level.

alpha (α): The specified maximum probability of a *Type I decision error*, i.e., the maximum probability of rejecting the *null hypothesis* when it is true. *Alpha* is also referred to as the *size of the test*. *Alpha* reflects the amount of evidence the *decision maker* would like to see before abandoning the *null hypothesis*.

alpha particle: A positively charged particle emitted by some radioactive materials undergoing *radioactive decay*.

alternative hypothesis (H_a): See *hypothesis*.

area: A general term referring to any portion of a *site*, up to and including the entire *site*.

area of elevated activity: An *area* over which *residual radioactivity* exceeds a specified value $DCGL_{EMC}$.

area factor (A_m): A factor used to adjust $DCGL_W$ to estimate $DCGL_{EMC}$ and the *minimum detectable concentration* for scanning surveys in *Class 1* survey units —

$DCGL_{EMC} = (DCGL_W)(A_m)$. A_m is the magnitude by which the *residual radioactivity* in a small *area of elevated activity* can exceed the $DCGL_W$ while maintaining compliance with the *release criterion*.

arithmetic mean: The average value obtained when the sum of individual values is divided by the number of values.

arithmetic standard deviation: A statistic used to quantify the variability of a set of data. It is calculated in the following manner: (1) subtracting the arithmetic mean from each data value individually; (2) squaring the differences; (3) summing the squares of the differences; (4) dividing the sum of the squared differences by the total number of data values less one; and (5) taking the square root of the quotient. The calculation process produces the Root Mean Square Deviation (RMSD).

background radiation: Radiation from cosmic sources; *naturally occurring radioactive material*, including radon (except as a decay product of *source* or *special nuclear material*); and global fallout as it exists in the environment from the testing of nuclear explosive devices or from nuclear accidents which contribute to *background radiation* and are not under the control of the licensee. *Background radiation* does not include radiation from *source*, *byproduct*, or *special nuclear materials* regulated by the NRC.

becquerel (Bq): The International System (SI) unit of activity equal to one nuclear transformation (disintegration) per second. *1 Bq = 2.7x10^{-11} Curies (Ci) = 27.03* picocuries (pCi).

beta (β): The probability of a *Type II decision error*, i.e., the probability of accepting the null hypothesis when it is false. The complement of *beta* $(1-\beta)$ is referred to as the *power* of the test.

beta particle: An electron emitted from the nucleus during *radioactive decay*.

bias: The systematic or persistent distortion of a measurement process which causes errors in one direction.

biased sample or measurement: See *judgment sample* or *measurement*.

byproduct material: Any radioactive material (except *special nuclear material*) yielded in or made radioactive by exposure to the radiation incident to the process of producing or utilizing *special nuclear material*.

CEDE (committed effective dose equivalent): The effective *dose equivalent* is the summation of the products of the dose equivalent received by specified tissues of the body and a tissue-specific weighting factor. It is a risk-equivalent value, expressed in *Sv* or *rem*, that can be used to estimate the health effects on an exposed individual. See *TEDE*.

characterization survey: A type of *survey* that includes facility or *site* sampling, monitoring, and analysis activities to determine the extent and nature of contamination. *Characterization* *surveys* provide the basis for acquiring necessary technical information to develop, analyze, and select appropriate *cleanup* techniques.

Class 1 area: Areas containing locations in which, prior to remediation, the concentrations of residual radioactivity may have exceeded the $DCGL_W$.

Class 1 survey: A type of *final status survey* that applies to *areas* with the highest potential for contamination. Class 1 surveys require (1) sufficient measurements on a systematic grid to meet the desired error rates for the statistical hypothesis tests; (2) scanning over 100% of the survey unit; (3) scanning MDC at or below the $DCGL_{EMC}$.

Class 2 area: Areas containing no locations where, prior to remediation, the concentrations of residual radioactivity may have exceeded the $DCGL_W$.

Class 2 survey: A type of *final status survey* that require (1) sufficient measurements on a systematic grid to meet the desired error rates for the statistical hypothesis tests; (2) judgmental scanning of a portion, up to 100%, of the survey unit.

Class 3 area: Areas with a low probability of containing any locations with residual radioactivity.

Class 3 survey: A type of *final status survey* that require (1) sufficient measurements at randomly chosen locations to meet the desired error rates for the statistical hypothesis tests; (2) judgmental scanning over less than 10% of the survey unit.

classification: The act or result of separating *areas* or *survey units* into one of three designated classes: *Class 1 area*, *Class 2 area*, or *Class 3 area*.

composite sample: A sample formed by collecting several samples and combining them (or selected portions of them) into a new sample which is then thoroughly mixed.

confirmatory survey: A type of *survey* that includes limited independent (third-party) measurements, sampling, and analyses to verify the findings of a *final status survey*.

contamination: The presence of *residual radioactivity* in excess of levels which are acceptable for release of a *site* or facility.

criterion: See *release criterion*.

critical group: The group of individuals reasonably expected to receive the greatest exposure to *residual radioactivity* for any applicable set of circumstances.

critical level (L_c): A fixed value of the *test statistic* corresponding to a given probability level, as determined from the sampling distribution of the *test statistic*. L_c is the level, in counts, at which there is a statistical probability (with a predetermined confidence) of incorrectly identifying a background value as greater than background.

curie (Ci): The customary unit of radioactivity. One *curie* (Ci) is equal to 37 billion disintegrations per second (3.7×10^{10} dps $= 3.7 \times 10^{10}$ Bq), which is approximately equal to the decay of one gram of ^{226}Ra. Fractions of a *curie*, e.g. picocurie (pCi) or 10^{-12} Ci and microcurie (μCi) or 10^{-6} Ci, are levels typically encountered in *decommissioning*.

DQA (Data Quality Assessment): The scientific and statistical evaluation of data to determine if the data are of the right type, quality, and quantity to support their intended use.

DQOs (Data Quality Objectives): Qualitative and quantitative statements that clarify study objectives, define the appropriate type of data, and specify levels of potential decision errors that will be used as the basis for establishing the quality and quantity of data needed to support decisions.

DCGL (derived concentration guideline level): A derived, radionuclide-specific activity concentration within a *survey unit* corresponding to the *release criterion*. The *DCGL* is based on the spatial distribution of the contaminant and hence is derived differently for the Wilcoxon test ($DCGL_W$) and the *Elevated Measurement Comparison* ($DCGL_{EMC}$). The $DCGL_W$ is derived assuming that residual radioactivity is uniformly distributed over a wide area, i.e. the entire survey unit. This can often be the default DCGL provided by an exposure pathway model. The $DCGL_{EMC}$ is derived assuming that residual radioactivity is concentrated in a much smaller area, i.e. in only a small percentage of the entire survey unit.

decay: See *radioactive decay*.

decision maker: The person, team, board, or committee responsible for the final decision regarding disposition of the *survey unit*.

decommission: To remove a facility or *site* safely from service and reduce *residual radioactivity* to a level that permits (1) release of the property for unrestricted use and termination of the *license* or (2) release of the property under restricted conditions and termination of the *license*.

decommissioning: The process of removing a facility or *site* from operation, followed by *decontamination* and license termination.

decontamination: The removal of radiological contaminants from, or their neutralization on, a person, object or area to within established levels. *Decontamination* is sometimes used interchangeably with *remediation*, remedial action, and *cleanup*.

delta (δ): The amount that the distribution of measurements for a *survey unit* is shifted to the right of the distribution of measurements of the *reference area*.

delta (Δ): The width of the *gray region*. Δ divided by the *arithmetic standard deviation* of the measurements, σ, is the *relative shift*, Δ/σ, expressed in multiples of standard deviations. See *relative shift, gray region*.

derived guideline(s): A level or levels of radioactivity presented in terms of ambient radiation, surface activity level(s), and soil activity concentration(s). Derived guidelines are derived from activity/dose relationships through various *exposure pathway* scenarios. See *DCGL*.

detection sensitivity: The ability to identify the presence of radiation or *radioactivity*.

direct measurement: Radioactivity measurement obtained by placing the detector against the surface or in the media being surveyed. The resulting radioactivity level is read out directly.

dose commitment: The dose that an organ or tissue would receive during a specified period of time (e.g., 50 or 70 years) as a result of intake (as by ingestion or inhalation) of one or more radionuclides from a given release.

dose equivalent (dose): A quantity that expresses all radiations on a common scale for calculating the effective absorbed dose. This quantity is the product of absorbed dose (rads) multiplied by a quality factor and any other modifying factors. Dose is measured in *Sv* or *rem*.

epsilon (ϵ): A fraction of a survey unit that has not been remediated to the reference-based cleanup standard. ϵ is used in the Quantile test.

elevated area: See *area of elevated activity*.

elevated measurement: A measurement that exceeds a specified value, the $DCGL_{EMC}$.

Elevated Measurement Comparison (EMC): This comparison is used in conjunction with the Wilcoxon test to determine if there are any measurements that exceed a specified value $DCGL_{EMC}$.

exposure pathway: The route by which radioactivity travels through the environment to eventually cause a radiation exposure to a person or group.

exposure rate: The amount of ionization produced per unit time in air by X-rays or gamma rays. The unit of exposure rate is roentgens/hour (R/h); for decommissioning activities the typical units are microroentgens per hour (μR/h), i.e. 10^{-6} R/h.

external radiation: Radiation from a source outside the body.

final status survey: Measurements and sampling to describe the radiological conditions of a survey unit, following completion of decontamination activities (if any).

gamma (γ) radiation: Penetrating high-energy, short-wavelength electromagnetic radiation (similar to X-rays) emitted during *radioactive decay*. Gamma-rays are very penetrating and require dense materials (such as lead or uranium) for shielding.

graded approach: An approach to data collection and interpretation that places the greatest survey efforts on areas that have, or had, the highest potential for *residual radioactivity*.

gray region: A range of values below the appropriate *DCGL* for a *site* in which decision errors carry relatively less risk or economic consequence.

grid: A network of intersecting and vertical lines used for the purpose of identification of exact locations. See *reference coordinate system, sampling grid*.

grid area: See *sampling grid area*.

grid block: A square defined by two adjacent vertical and two adjacent horizontal *reference coordinate* lines.

half-life ($t_{1/2}$): The time required for one-half of the atoms present to disintegrate.

Historical Site Assessment (HSA): A detailed investigation to collect existing information, primarily historical information, on a *site* and its surroundings.

hot measurement: See *elevated measurement*.

hot spot: See *area of elevated activity*.

hypothesis: An assumption about a property or characteristic of a set of data under study. The goal of statistical inference is to decide which of two complementary hypotheses is likely to be true. The *null hypothesis* describes what is assumed to be the true state of nature and the *alternative hypothesis* describes the opposite situation.

impacted area: Areas with some potential for residual contamination .

indistinguishable from background: The detectable concentration distribution of a radionuclide is not statistically different from the background concentration distribution of that radionuclide in the vicinity of the site or, in the case of structures, in similar materials using adequate measurement technology, survey, and statistical techniques.

inventory: Total residual quantity of formerly licensed radioactive material at a site.

investigation level: A radionuclide specific level of radioactivity that results in additional investigation when it is exceeded, to determine if the survey unit was properly classified. See *action level, elevated measurement*.

judgment sample or measurement: A sample or measurement taken at a location where radiation levels or other site characteristics are expected to be unusual, based on the judgment and/or experience of a skilled investigator. Also called authoritative or biased. Samples or measurements that are *not* of this type are considered representative of the site being studied.

k: When conducting the Quantile test, k is the number of measurements from the survey unit that are among the r largest measurements of the combined set of reference area and cleanup unit measurements.

less-than data: Measurements that are less than the *minimum detectable concentration.*

license: A license issued under the regulations in Parts 30 through 35, 39, 40, 60, 61, 70 or Part 72 of 10 CFR Chapter I.

licensee: The holder of a *license.*

license termination: Discontinuation of a *license,* the eventual conclusion to *decommissioning.*

lower bound of the gray region (LBGR): The minimum value of the gray region. The width of the *gray region (DCGL−LBGR)* is also referred to as the shift, Δ.

lower limit of detection (L_D): The smallest amount of radiation or radioactivity that statistically yields a net result above the method background. The critical level, L_C, is the value set for deciding that radioactivity is detected with a specified Type I error rate. The detection limit, L_D, is the level at which the power to detect net radioactivity is set.

m: The number of measurements required from the reference area to conduct a statistical test with specified *Type I* and *Type II* error rates.

minimum detectable concentration (MDC): The *a priori* activity level that a specific instrument and technique can be expected to detect a specified percentage of the time. When stating the detection capability of an instrument, this value should be used. The *MDC* is the detection limit, L_D, multiplied by an appropriate conversion factor to give units of activity.

missing or unusable data: Data (measurements) that are mislabeled, lost, or do not meet quality control standards. *Less-than data* are not considered to be missing or unusable data.

N: $N = m + n$, is the total number of measurements required from the reference area and a *survey unit* being compared with the *reference area.* See m and n.

n: Number of measurements required from a survey unit to conduct a statistical test that has specified *Type I* and *Type II* error rates.

naturally occurring radionuclides: Radionuclides and their associated progeny produced during the formation of the earth or by interactions of terrestrial matter with cosmic rays.

non-detect: A measurement below the *critical level, L_C.*

non-impacted area: Areas in which there is no reasonable possibility of residual contamination.

nonparametric test: A test based on relatively few assumptions about the exact form of the underlying probability distributions of the measurements. As a consequence, nonparametric tests are generally valid for a fairly broad class of distributions. The *Wilcoxon Rank Sum test* and the *Sign test* are examples of nonparametric tests.

normal (Gaussian) distribution: A family of bell-shaped distributions described by the mean and variance.

outlier: Measurements that are unusually large relative to the bulk of the measurements in the data set.

Pitman efficiency: A measure of performance for statistical tests. It is equal to the reciprocal of the ratio of the sample sizes required by each of two tests to achieve the same power, as these sample sizes become large.

power (1-β): The probability of rejecting the *null hypothesis* when it is false. The power is equal to one minus the *Type II* error rate, i.e. $(1-\beta)$.

precision: A measure of mutual agreement among individual measurements, usually under prescribed similar conditions, expressed generally in terms of the *arithmetic standard deviation*.

quality assurance (QA): An integrated system of management activities involving planning, implementation, assessment, reporting, and quality improvement to ensure that a process, item, or service is of the type and quality needed and expected by the client.

Quality Assurance Project Plan (QAPP): A formal document describing in comprehensive detail the necessary *QA*, *QC*, and other technical activities that must be implemented to ensure that the results of the work performed satisfies the stated performance criteria.

quality control (QC): The overall system of technical activities that measure the attributes and performance of a process, item, or service against defined standards to verify that they meet the stated requirements established by the client. *QC* includes operational techniques and activities that are used to fulfill requirements for *quality*.

Quantile Test: A nonparametric test that looks at only the *r* largest measurements of the N combined reference area and survey unit measurements. If a sufficiently large number of these *r* measurements are from the survey unit, then the test indicates the survey unit has not attained the reference-based cleanup standard.

R_A: The acceptable level of risk associated with not detecting an *area of elevated activity* of area A_{min}.

radiation (ionizing radiation): Alpha particles, beta particles, gamma rays, x-rays, neutrons, high-speed electrons, high-speed protons, and other particles capable of producing ions. Radiation, as used in this document, does not include non-ionizing radiation, such as radio- or microwaves, or visible, infrared, or ultraviolet light.

radioactive decay: The spontaneous transformation of an unstable atom into one or more different nuclides accompanied by either the emission of energy and/or particles from the nucleus, nuclear capture or ejection of orbital electrons, or fission. Unstable atoms decay into a more stable state, eventually reaching a form that does not decay further or has a very long *half-life*.

radioactivity: The mean number of nuclear transformations occurring in a given quantity of radioactive material per unit time. The International System (SI) unit of radioactivity is the *Becquerel (Bq)*. The customary unit is the *Curie (Ci)*.

radiological survey: Measurements of radiation levels associated with a *site* together with appropriate documentation and data evaluation.

radionuclide: An unstable nuclide that undergoes *radioactive decay*.

random error: The deviation of an observed value from the true value is called the error of observation. If the error of observation behaves like a random variable—i.e., its value occurs as though chosen at random from a probability distribution of such errors—it is called a *random error*. See *systematic error*.

readily removable: A qualitative statement of the extent to which a radionuclide can be removed from a surface or medium using non-destructive, common, housekeeping techniques (e.g., washing with moderate amounts of detergent and water) that do not generate large volumes of radioactive waste requiring subsequent disposal or produce chemical wastes that are expected to adversely affect public health or the environment.

reference area: Geographical *area* from which representative reference measurements are performed for comparison with measurements performed in specific *survey units* at remediation site. A site radiological *reference area* (background area) is defined as an area that has similar physical, chemical, radiological, and biological characteristics as the site area being remediated, but which has not been contaminated by site activities. The distribution and concentration of *background radiation* in the *reference area* should be the same as that which would be expected on the *site* if that *site* had never been contaminated. More than one *reference area* may be necessary may be necessary for valid comparisons if a *site* exhibits considerable physical, chemical, radiological, or biological variability.

reference coordinate system: A *grid* of intersecting lines referenced to a fixed site location or benchmark. Typically the lines are arranged in a perpendicular pattern dividing the survey location into squares or blocks of equal areas. Other patterns include three-dimensional and polar coordinate systems.

reference grid: A network of parallel horizontal and vertical lines forming squares on a map that may be overlaid on a property parcel for the purpose of identification of exact locations. See *reference coordinate system*.

reference region: The geographical region from which *reference areas* will be selected for comparison with *survey units*.

relative shift (Δ/σ): Δ divided by σ, the *standard deviation* of the measurements, is the *relative shift* expressed in multiples of standard deviations. See *delta*.

release criteria: The criteria for license termination given in 10 CFR 20.1402 and 10 CFR 20.1403, expressed in terms of total effective dose equivalent (TEDE).

release criterion: A regulatory limit expressed in terms of dose or risk.

rem (roentgen equivalent man): The conventional unit of *dose equivalent*. The corresponding International System (SI) unit is the *Sievert (Sv)*: 1 Sv = 100 rem.

remediation: The process and associated activities resulting in removal of contamination from a site. Remediation is sometimes used interchangeably with the terms remedial action, response action, or *decontamination*.

remediation control survey: A type of survey that includes monitoring the progress of remedial action by real time measurement of areas being decontaminated to determine whether or not efforts are effective and to guide further *decontamination* activities.

removable activity: Surface activity that can be *readily removed* and collected for measurement by wiping the surface with moderate pressure.

residual radioactivity: Radioactivity in structures, materials, soils, groundwater, and other media at a site resulting from activities under the licensee's control. This includes radioactivity from all licensed and unlicensed sources used by the licensee, but excludes background radioactivity. It also includes radioactive materials remaining at the site as a result of routine or accidental releases of radioactive material at the site and previous burials at the site, even if those burials were made in accordance with the provisions of 10 CFR Part 20.

restricted use: A designation following *remediation* requiring radiological controls at a formerly licensed *site*.

sample: A part or selection from a medium located in a *survey unit* or *reference area* that represents the quality or quantity of a given parameter or nature of the whole area or unit; a portion serving as a specimen.

sampling grid: The pattern of points formed by the locations chosen for *systematic sampling*.

sampling grid area: The area bounded by adjacent sampling locations on a systematic *sampling grid*. If the linear distance between samples is L, then the grid area is L^2 for a square sampling grid and $0.866L^2$ for a triangular sampling grid.

scanning: An evaluation technique performed by moving a detection device over a surface at a specified speed and distance above the surface to detect elevated levels of radiation.

scoping survey: A type of *survey* that is conducted to identify (1) radionuclide contaminants, (2) relative radionuclide ratios, and (3) general levels and extent of contamination.

shape parameter (S): For an elliptical area of elevated activity, the ratio of the semi-minor axis length to the semi-major axis length is the shape parameter. For a circle, the shape parameter is one. A small shape parameter corresponds to a flat ellipse.

Sievert (Sv): The special name for the International System (SI) unit of *dose equivalent*. 1 Sv = 100 *rem* = 1 Joule per kilogram.

Sign test: A *nonparametric* statistical test used to demonstrate compliance with the release criterion when the radionuclide of interest is not present in background and the distribution of data is not symmetric. See also *Wilcoxon Rank Sum test*.

site: Any installation, facility, or discrete, physically separate parcel of land, or any building or structure or portion thereof, that is being considered for release.

size (of a test): See *alpha*.

soil: The top layer of the earth's surface, consisting of rock and mineral particles mixed with organic matter. A particular kind of earth or ground—e.g., sandy soil.

soil activity (soil concentration): The level of radioactivity present in soil and expressed in units of activity per soil mass (typically Bq/kg or pCi/g).

source material: Uranium and/or thorium other than that classified as *special nuclear material*.

source term: All residual radioactivity remaining at the *site*, including material released during normal operations, during inadvertent releases or accidents, and includes radioactive materials which may have been buried at the site in accordance with 10 CFR Part 20.

special nuclear material: Plutonium, ^{233}U, and uranium enriched in ^{235}U; material capable of undergoing a fission reaction.

square sampling grid: A systematic grid of sampling locations that is arranged in a square pattern. See *sampling grid*.

standard normal distribution: A *normal (Gaussian) distribution* with mean zero and variance one.

standard operating procedure (SOP): A written document that details the method for an operation, analysis, or action with thoroughly prescribed techniques and steps, and that is officially approved as the method for performing certain routine or repetitive tasks.

subsurface soil sample: A soil sample taken deeper than 15 cm below the soil surface.

surface contamination: *Residual radioactivity* found on building or equipment surfaces and expressed in units of activity per surface area (Bq/m^2 or dpm/100 cm^2).

surface soil sample: A soil sample taken from the first 15 cm of surface soil.

surrogate: Radionuclide A is a *surrogate* for radionuclide B if there is an established ratio between their concentrations in a survey unit. The concentration of radionuclide B can then be inferred from the measured concentration of radionuclide A.

survey: A systematic evaluation and documentation of radiological measurements with a correctly calibrated instrument or instruments that meet the sensitivity required by the objective of the evaluation.

survey plan: A plan for determining the radiological characteristics of a *site*.

survey unit: A geographical area of specified size and shape at a remediated site for which a separate decision will be made whether the unit attains the site-specific reference-based cleanup standard for the designated pollution parameter. *Survey units* are generally formed by grouping contiguous site areas with a similar use history and the same classification of contamination potential. Survey units are established to facilitate the survey process and the statistical analysis of survey data.

systematic error: An error of observation based on system faults which are biased in one or more ways, e.g., tending to be on one side of the true value more than the other.

systematic sampling: Taking measurements or samples at locations according to a spatial pattern. See *triangular sampling grid*, *square sampling grid*.

tandem testing: Two or more statistical tests conducted using the same data set.

TEDE (total effective dose equivalent): The effective dose equivalent is the summation of the products of the *dose equivalent* received by specified tissues of the body and a tissue-specific weighting factor. It is a risk-equivalent value, expressed in *Sv* or *rem*, that can be used to estimate the health-effects on an exposed individual. When calculating *TEDE*, the licensee shall base estimates on the greatest annual *TEDE* dose expected within the first 1000 years after decommissioning. Estimates must be substantiated using actual measurements to the maximum extent practical. See *CEDE*.

test statistic: A function of the measurements (or their ranks) that has a known distribution if the *null hypothesis* is true. This is compared to the *critical level* to determine if the *null hypothesis* should be rejected.

tied measurements: Two or more measurements that have the same value.

triangular sampling grid: A grid of sampling locations that is arranged in a triangular pattern. See *sampling grid*.

two-sample t-test: A parametric statistical test used in place of the *Wilcoxon Rank Sum (WRS) test* if the *reference area* and *survey unit* measurements are known to be *normally (Gaussian) distributed* and there are no *less-than measurements* in either data set.

Type I decision error: A decision error that occurs when the *null hypothesis* is rejected when it is true. The probability of making a *Type I decision error* is called *alpha* (α).

Type II decision error: A decision error that occurs when the *null hypothesis* is accepted when it is false. The probability of making a *Type II decision error* is called *beta* (β).

unaffected area: An impacted area that is expected to contain little, if any, residual radioactivity, based on a knowledge of site history and previous survey information. This term was previously used for areas designated *Class 3*.

unity rule (mixture rule): A rule applied when more than one radionuclide is present at a concentration that is distinguishable from background and where a single concentration comparison does not apply. In this case the mixture of radionuclides is compared against default concentrations by applying the unity rule. This is accomplished by determining (1) the ratio between the concentration of each radionuclide in the mixture and (2) the concentration for that radionuclide in an appropriate listing of default values. The sum of the ratios for all radionuclides in the mixture should not exceed 1.

unrestricted area: Any *area* where access is not controlled by a *licensee* for purposes of protection of individuals from exposure to radiation and radioactive materials—including areas used for residential purposes.

unrestricted release: Release of a *site* from regulatory control without requirements for future radiological restrictions. Also known as unrestricted use.

Wilcoxon Rank Sum (WRS) test: A *nonparametric* statistical test used to determine compliance with the *release criterion* when the radionuclide of concern is present in background. See also *Sign test*.

z_ϕ: the ϕth quantile of a standard normal distribution. Also called the 100ϕ percentile. The probability of observing a value less than z_ϕ is equal to ϕ. See *standard normal distribution*.

16 BIBLIOGRAPHY

The material in this bibliography has been helpful in the development of this document.

Abramowitz, M. , and I.A. Stegun (eds.), *Handbook of Mathematical Functions.* New York: Dover. 1972.

Andrews, F.C., "Asymptotic Behavior of Some Rank Tests for Analysis of Variance," Annals of Mathematical Statistics, 25:724–736. 1954.

ANL/EAD/LD-2 (Argonne National Laboratory/Environmental Assessments Division). "Manual for Implementing Residual Radioactive Material Guidelines Using RESRAD, Version 5." Washington, D.C.: Department of Energy. 1993.

ANL/EAD/LD-3 (Argonne National Laboratory/Environmental Assessments Division).. "RESRAD-BUILD: A Computer Model for Analyzing the Radiological Doses Resulting From the Remediation and Occupancy of Buildings Contaminated With Radioactive Material." Washington, D.C.: Department of Energy. 1994.

ANSI/ASQC E-1994. "Specifications and Guidelines for Quality Systems for Environmental Data Collection and Environmental Technology Programs". Milwaukee, Wisconsin: American National Standards Institute/American Society for Quality Control. 1994.

BEIR (National Academy of Sciences/National Research Council Committee on the Biological Effects of Ionizing Radiations). *Health Effects of Exposure to Low Levels of Ionizing Radiation, BEIR V.* Washington, D.C.: National Academy Press. 1990.

Box, G.E.P., *Biometrika*, 40, 318–335, 1953.

Brownlee, K.A., *Statistical Theory and Methodology in Science and Engineering.* New York: Wiley. 1960.

Chen, L., "Testing the Mean of Skewed Distributions." *Journal of the American Statistical Association.* 90, 430, 767–772. 1995.

Christensen, R. *Linear Models for Multivariate, Time Series and Spatial Data.* New York: Springer. 1991.

Clark, I. *Practical Geostatistics.* New York: Elsevier. 1979.
Conover, W.J. *Practical Nonparametric Statistics*, 2d edition. New York: Wiley. 1980 (2nd ed.)

Cox, F.M., and C.F. Guenther. "An Industry Survey of Current Lower Limits of Detection for Various Radionuclides," *Health Physics* 69(1):121–129. 1995.

Cressie, N.A.C. *Statistics for Spatial Data.* New York: Wiley. 1993.

BIBLIOGRAPHY

Currie, L.A. "Limits for Qualitative Detection and Quantitative Determination," *Analytical Chemistry* 40(3): 586–693. 1968.

Diggle, P.J. *Statistical Analysis of Spatial Point Patterns.* San Diego, Cal.: Academic Press. 1983.

DOE/EM-0142PDOE. *Decommissioning Handbook.* Washington, D.C.: Department of Energy. March 1994.

DOE/RL-92-24. *Hanford Site Soil Background.* Richland, Wash.: Department of Energy. 1992.

DOE/RL-94-72. *Hanford Site Background Data Applications Guide: Part 1, Soil.* Washington, D.C.: Department of Energy. June 1994.

Eberhardt, K.R. "Survey Sampling Methods," Chapter 9 in H.M. Wadsworth (ed.), *Handbook of Statistical Methods for Scientists and Engineers,*" p. 9.20. New York: McGraw-Hill. 1990.

Efron, B., and R.J. Tibshirani. *An Introduction to the Bootstrap.* New York: Chapman & Hall. 1993.

EPA QA/G-4. *Guidance for the Data Quality Objectives Process, Final.* Washington, D.C.:Environmental Protection Agency. 1994.

EPA QA/G-5. *Guidance on Quality Assurance Project Plans.* Washington, D.C.:Environmental Protection Agency. 1996.

EPA QA/G-9. *Guidance for Data Quality Assessment, QA96 Version.* Washington, D.C.: Environmental Protection Agency. July 1996.
EPA QA/G-9. *Guidance for Data Quality Assessment, External Working Draft.* Washington, D.C.: Environmental Protection Agency. March 27, 1995.

EPA QA/R-5. *EPA Requirements for Quality Assurance Project Plans for Environmental Data Operations.* Washington, D.C.:Environmental Protection Agency. 1994.

EPA 230-R-92-14. *Methods for Evaluating the Attainment of Cleanup Standards, Volume 2: Groundwater.* Washington, D.C.: Environmental Protection Agency, Statistical Policy Branch. 1992.

EPA 230-R-94-004. *Statistical Methods for Evaluating the Attainment of Cleanup Standards, Volume 3: Reference Based Standards for Soils and Solid Media.* Washington, D.C.: Environmental Protection Agency, Office of Policy, Planning, and Evaluation. 1994.

EPA PB92- 963373. *Supplemental Guidance to Rags [Risk Assessment Guidance For Superfund]: Calculating the Concentration Term --* Washington, DC: Environmental Protection Agency, Office of Solid Waste and Emergency Response, 1992.

EPA 230/02-89-042. *Methods for Evaluating the Attainment of Cleanup Standards, Volume 1: Soils and Solid Media.* Washington, D.C.: Environmental Protection Agency, Statistical Policy Branch. 1989.

EPAv/S40/1-89/002. *Risk Assessment Guidance for Superfund, Volume I: Human Health Evaluation Manual (Part A), Interim Final.* Washington, D.C.: Environmental Protection Agency, Office of Emergency and Remedial Response. 1989.

EPA 540/G-87/003. *Data Quality Objectives for Remedial Response Activities: Development Process.* Washington, D.C.: Environmental Protection Agency. 1987.

EPA 520/1-80-012. *Upgrading Environmental Radiation Data.* Washington, D.C.: Environmental Protection Agency. August 1980.

EPA 540/G-87/004. *Data Quality Objectives for Remedial Response Activities: Example Scenario RI/FS Activities at a Site With Contaminated Soils and Ground Water.* Washington, D.C.: Environmental Protection Agency. 1987.

EPA 600/4-84-043. "Soil Sampling Quality Assurance User's Guide," 1st ed. Washington, D.C.: Environmental Protection Agency. 1984.
EPA 600/4-90/013. *A Rationale for the Assessment of Errors in the Sampling of Soils.* Washington, D.C.: Environmental Protection Agency. 1990.

Federal Register, July 21, 1997. 62 FR 139, pp. 39057–39092.

Gilbert, R.O. *Statistical Methods for Environmental Pollution Monitoring.* New York: Van Nostrand Reinhold. 1987.

Hastings, C., Jr. *Approximations for Digital Computers.* Princeton, N.J.: Princeton University Press. 1955.

Hochberg, Y., and A.C. Tamhane. *Multiple Comparison Procedures.* New York: Wiley. 1987.

Johnson, R.A., S. Verrill, and D.H. Moore II. "Two-Sample Rank Tests for Detecting Changes That Occur in a Small Proportion of the Treated Population," *Biometrics* 43:641–655. 1987.

Land, C.E. "Hypothesis Tests and Interval Estimates" in E.L. Crow and K. Shimizu (eds.), *Lognormal Distributions: Theory and Practice.* New York: Marcel Dekker, Inc. 1988.

Lehmann, E.L., and H.J.M. D'Abrera. *Nonparametrics:Statistical Methods Based on Ranks.* San Francisco: Holden-Day, Inc. 1975.

Liggett, W. "Detecting Elevated Contamination by Comparisons With Background," pp. 119–128 in G.E. Schweitzer and J.A. Santoluctto (eds.), *Environmental Sampling for Hazardous Wastes,* ACS Symposium Series 267. Washington, D.C.: American Chemical Society. 1984.

BIBLIOGRAPHY

Ling, R.F., and J.W. Pratt. "The Accuracy of Peizer Approximations to the Hypergeometric Distribution, With Comparisons to Some Other Approximations," *Journal of the American Statistical Association* 79 (385) 49–60. 1984.

Microsoft Excel for Windows, Version 5.0. Redmond, WA: Microsoft Corporation. 1993.

Neptune, O., E.P. Brantley, M.J. Messner, and D.I. Michael. "Quantitative Decision Making in Superfund: A Data Quality Objectives Case Study," *Hazardous Materials Control* 3(3):18–27. 1990.

Noether, G.E. "Sample Size Determination for Some Common Nonparametric Tests," *Journal of the American Statistical Association* 82:645–647. 1987.

Norton, R.M. "Pocket-Calculator Approximation for Areas Under the Standard Normal Curve," *American Statistician* 43:24–26. 1989.

Nuclear Regulatory Commission. "Draft Branch Technical Position on Site Characterization for Decommissioning." Washington, D.C. November 1994.

Nuclear Regulatory Commission. "Guidelines for Decontamination of Facilities and Equipment Prior to Release for Unrestricted Use or Termination of Licenses for Byproduct, Source, or Special Nuclear Materials." Washington, D.C. May 1987.

NUREG-0586. "Final Generic Environmental Impact Statement on Decommissioning of Nuclear Facilities." Washington, D.C.: Nuclear Regulatory Commission. 1988.

NUREG-1496 [a]. "Generic Environmental Impact Statement in Support of Rulemaking on Radiological Criteria for Decommissioning of NRC-Licensed Facilities," Volume 1— Main Report (Draft Report for Comment). Washington, D.C.: Nuclear Regulatory Commission. August 1994.

NUREG-1496 [b]. "Generic Environmental Impact Statement in Support of Rulemaking on Radiological Criteria for Decommissioning of NRC-Licensed Facilities," Volume 2—Appendices (Draft Report for Comment). Washington, D.C.: Nuclear Regulatory Commission. August 1994.

NUREG-1500. "Working Draft Regulatory Guide on Release Criteria for Decommissioning: NRC Staff's Draft for Comment." Washington, D.C.: Nuclear Regulatory Commission. August 1994.

NUREG-1501. "Background as a Residual Radioactivity Criterion for Decommissioning" (Draft Report). Washington, D.C.: Nuclear Regulatory Commission. August 1994.

NUREG-1505 Nuclear Regulatory Commission (NRC). 1997a. "A Nonparametric Statistical Methodology for the Design and Analysis of Final Status Decommissioning Surveys" (Draft Report for Comment). Washington, D.C.: Nuclear Regulatory Commission. August 1995.

NUREG-1506. "Measurement Methods for Radiological Surveys in Support of New Decommissioning Criteria" (Draft Report for Comment). Washington, D.C.: Nuclear Regulatory Commission. 1995.

NUREG/CR-1507. "Minimum Detectable Concentrations with Typical Radiation Survey Instruments for Various Contaminants and Field Conditions" (Draft Report for Comment). Washington, D.C.: Nuclear Regulatory Commission. 1995.

NUREG/CR-4007. "Lower Limit of Detection: Definition and Elaboration of a Proposed Position for Radiological Effluent and Environmental Measurements." Washington, D.C.: Nuclear Regulatory Commission, 1984.

NUREG/CR-5512. "Residual Radioactive Contamination From Decommissioning," Volume 1 (Draft). Washington, D.C.: Nuclear Regulatory Commission. January 1990.

NUREG/CR-5849. "Manual for Conducting Radiological Surveys in Support of License Termination." Washington, D.C.: Nuclear Regulatory Commission. June 1992.

ORNL/TM-12774. "ELIPGRID-PC: A PC Program for Calculating Hot Spot Probabilities." Grand Junction, Co.: Oak Ridge National Laboratory. 1994.

PNL-7409. "Statistical Methods for Evaluating the Attainment of Cleanup Standards," Volume 3: "Reference-Based Standards for Soils and Solid Media." Richland, Wash.: Pacific Northwest Laboratory. 1992.

PNL-8989. " Comparing Statistical Tests for Detecting Soil Contamination Greater Than Background." Richland, Wash.: Pacific Northwest Laboratory. 1993.

Regulatory Guide 1.86. "Termination of Operating Licenses for Nuclear Reactors." Washington, D.C.: Nuclear Regulatory Commission. June 1974.

Regulatory Guide 4.15. "Quality Assurance for Radiological Monitoring Programs — Effluent Streams and the Environment." Washington, D.C.: Nuclear Regulatory Commission. February 1979.

Ripley, B.D. *Spatial Statistics.* New York: Wiley. 1981.

Rohatgi, V.K. *An Introduction to Probability Theory and Mathematical Statistics.* New York: Wiley. 1976.

Ryti, R.T., and D. Neptune. "Planning Issues for Superfund Site Remediation." *Hazardous Materials Control* 4(6):47–53. 1991.

Scott, D.W, "On Optimal and Data-based Histograms," *Biometrika* 66, 605–610. 1979.

Singer, D.A. "ELIPGRID: A FORTRAN IV Program for Calculating the Probability of Success in Locating Elliptical Targets With Square, Rectangular and Hexagonal Grids," *Geocom Programs* 4:1–19. 1972.

Singer, D.A. "Relative Efficiencies of Square and Triangular Grids in the Search for Elliptically Shaped Resource Targets," *Journal of Research of the U.S. Geological Survey* 3(2):163–167. 1975.

Snedecor, G.W., and W.G. Cochran. *Sampling Techniques* (7th edition). New York: Wiley. 1980.

Sprent, P. *Applied Nonparametric Statistics*, 2d ed. New York: Chapman and Hall. 1993. 1993.

Spurrier, J.D. and J. E. Hewett. "Two-Stage Wilcoxon Tests of Hypotheses," *Journal of the American Statistical Association*, 71(356):982–987. 1976.

Taylor, J. K. *Statistical Techniques for Data Analysis*. Chelsea, Mich.: Lewis Publishers. 1990.

Title 10 ("Energy"), *Code of Federal Regulations*, Part 20 ("Standards for Protection Against Radiation."). U.S. Nuclear Regulatory Commission (updated periodically).

Title 10 ("Energy"), *Code of Federal Regulations*, Part 30 ("Rules of General Applicability to Domestic Licensing of Byproducts and Material."). U.S. Nuclear Regulatory Commission (updated periodically).

Title 10 ("Energy"), *Code of Federal Regulations*, Part 40 ("Domestic Licensing of Source Material."). U.S. Nuclear Regulatory Commission (updated periodically).

Title 10 ("Energy"), *Code of Federal Regulations*, Part 50 ("Domestic Licensing of Production and Utilization Facilities."). U.S. Nuclear Regulatory Commission (updated periodically).

Title 10 ("Energy"), *Code of Federal Regulations*, Part 70 ("Domestic Licensing of Special Nuclear Material."). U.S. Nuclear Regulatory Commission (updated periodically).

Title 10 ("Energy"), *Code of Federal Regulations*, Part 72 ("Licensing Requirements for the Independent Storage of Spent Nuclear Fuel and High-level Radioactive Waste"). U.S. Nuclear Regulatory Commission (updated periodically).

Thomson, G.W. *Biometrika*, 42, 268–269. 1955.

Upton, G., and B. Fingleton. *Spatial Data Analysis by Example*. New York: Wiley. 1985

Wadsworth, H.M. *Handbook of Statistical Methods for Scientists and Engineers*. New York: McGraw-Hill. 1990.

Zirschky, J., and R.O. Gilbert. "Detecting Hot Spots at Hazardous Waste Sites," *Chemical Engineering* 9:97–100. 1984.

APPENDIX: STATISTICAL TABLES

Table A.1 Cumulative Normal Distribution Function $\Phi(z)$

z	0.00	0.01	0.02	0.03	0.04	0.05	0.06	0.07	0.08	0.09
0.00	0.5000	0.5040	0.5080	0.5120	0.5160	0.5199	0.5239	0.5279	0.5319	0.5359
0.10	0.5398	0.5438	0.5478	0.5517	0.5557	0.5596	0.5636	0.5674	0.5714	0.5753
0.20	0.5793	0.5832	0.5871	0.5910	0.5948	0.5987	0.6026	0.6064	0.6103	0.6141
0.30	0.6179	0.6217	0.6255	0.6293	0.6331	0.6368	0.6406	0.6443	0.6480	0.6517
0.40	0.6554	0.6591	0.6628	0.6664	0.6700	0.6736	0.6772	0.6808	0.6844	0.6879
0.50	0.6915	0.6950	0.6985	0.7019	0.7054	0.7088	0.7123	0.7157	0.7190	0.7224
0.60	0.7257	0.7291	0.7324	0.7357	0.7389	0.7422	0.7454	0.7486	0.7517	0.7549
0.70	0.7580	0.7611	0.7642	0.7673	0.7704	0.7734	0.7764	0.7794	0.7823	0.7852
0.80	0.7881	0.7910	0.7939	0.7967	0.7995	0.8023	0.8051	0.8078	0.8106	0.8133
0.90	0.8159	0.8186	0.8212	0.8238	0.8264	0.8289	0.8315	0.8340	0.8365	0.8389
1.00	0.8413	0.8438	0.8461	0.8485	0.8508	0.8531	0.8554	0.8577	0.8599	0.8621
1.10	0.8643	0.8665	0.8686	0.8708	0.8729	0.8749	0.8770	0.8790	0.8810	0.8830
1.20	0.8849	0.8869	0.8888	0.8907	0.8925	0.8944	0.8962	0.8980	0.8997	0.9015
1.30	0.9032	0.9049	0.9066	0.9082	0.9099	0.9115	0.9131	0.9147	0.9162	0.9177
1.40	0.9192	0.9207	0.9222	0.9236	0.9251	0.9265	0.9279	0.9292	0.9306	0.9319
1.50	0.9332	0.9345	0.9357	0.9370	0.9382	0.9394	0.9406	0.9418	0.9429	0.9441
1.60	0.9452	0.9463	0.9474	0.9484	0.9495	0.9505	0.9515	0.9525	0.9535	0.9545
1.70	0.9554	0.9564	0.9573	0.9582	0.9591	0.9599	0.9608	0.9616	0.9625	0.9633
1.80	0.9641	0.9649	0.9656	0.9664	0.9671	0.9678	0.9686	0.9693	0.9699	0.9706
1.90	0.9713	0.9719	0.9726	0.9732	0.9738	0.9744	0.9750	0.9756	0.9761	0.9767
2.00	0.9772	0.9778	0.9783	0.9788	0.9793	0.9798	0.9803	0.9808	0.9812	0.9817
2.10	0.9821	0.9826	0.9830	0.9834	0.9838	0.9842	0.9846	0.9850	0.9854	0.9857
2.20	0.9861	0.9864	0.9868	0.9871	0.9875	0.9878	0.9881	0.9884	0.9887	0.9890
2.30	0.9893	0.9896	0.9898	0.9901	0.9904	0.9906	0.9909	0.9911	0.9913	0.9916
2.40	0.9918	0.9920	0.9922	0.9925	0.9927	0.9929	0.9931	0.9932	0.9934	0.9936
2.50	0.9938	0.9940	0.9941	0.9943	0.9945	0.9946	0.9948	0.9949	0.9951	0.9952
2.60	0.9953	0.9955	0.9956	0.9957	0.9959	0.9960	0.9961	0.9962	0.9963	0.9964
2.70	0.9965	0.9966	0.9967	0.9968	0.9969	0.9970	0.9971	0.9972	0.9973	0.9974
2.80	0.9974	0.9975	0.9976	0.9977	0.9977	0.9978	0.9979	0.9979	0.9980	0.9981
2.90	0.9981	0.9982	0.9982	0.9983	0.9984	0.9984	0.9985	0.9985	0.9986	0.9986
3.00	0.9987	0.9987	0.9987	0.9988	0.9988	0.9989	0.9989	0.9989	0.9990	0.9990
3.10	0.9990	0.9991	0.9991	0.9991	0.9992	0.9992	0.9992	0.9992	0.9993	0.9993
3.20	0.9993	0.9993	0.9994	0.9994	0.9994	0.9994	0.9994	0.9995	0.9995	0.9995
3.30	0.9995	0.9995	0.9995	0.9996	0.9996	0.9996	0.9996	0.9996	0.9996	0.9997
3.40	0.9997	0.9997	0.9997	0.9997	0.9997	0.9997	0.9997	0.9997	0.9997	0.9998

Negative values of z can be obtained from the relationship $\Phi(-z) = 1 - \Phi(z)$.

Table A.2a Sample Sizes for the Sign Test
(Number of measurements to be performed in each survey unit)

Δ/σ	0.01 0.01	0.01 0.025	0.01 0.05	0.01 0.1	0.01 0.25	0.025 0.025	0.025 0.05	0.025 0.1	0.025 0.25	0.05 0.05	0.05 0.1	0.05 0.25	0.1 0.1	0.1 0.25	0.25 0.25
0.10	4095	3476	2984	2463	1704	2907	2459	1989	1313	2048	1620	1018	1244	725	345
0.20	1035	879	754	623	431	735	622	503	333	518	410	258	315	184	88
0.30	468	398	341	282	195	333	281	227	150	234	185	117	143	83	40
0.40	270	230	197	162	113	192	162	131	87	136	107	68	82	48	23
0.50	178	152	130	107	75	126	107	87	58	89	71	45	54	33	16
0.60	129	110	94	77	54	92	77	63	42	65	52	33	40	23	11
0.70	99	83	72	59	41	70	59	48	33	50	40	26	30	18	9
0.80	80	68	58	48	34	57	48	39	26	40	32	21	24	15	8
0.90	66	57	48	40	28	47	40	33	22	34	27	17	21	12	6
1.00	57	48	41	34	24	40	34	28	18	29	23	15	18	11	5
1.10	50	42	36	30	21	35	30	24	17	26	21	14	16	10	5
1.20	45	38	33	27	20	32	27	22	15	23	18	12	15	9	5
1.30	41	35	30	26	17	29	24	21	14	21	17	11	14	8	4
1.40	38	33	28	23	16	27	23	18	12	20	16	10	12	8	4
1.50	35	30	27	22	15	26	22	17	12	18	15	10	11	8	4
1.60	34	29	24	21	15	24	21	17	11	17	14	9	11	6	4
1.70	33	28	24	20	14	23	20	16	11	17	14	9	10	6	4
1.80	32	27	23	20	14	22	20	16	11	16	12	9	10	6	4
1.90	30	26	22	18	14	22	18	15	10	16	12	9	10	6	4
2.00	29	26	22	18	12	21	18	15	10	15	12	8	10	6	3
2.50	28	23	21	17	12	20	17	14	10	15	11	8	9	5	3
3.00	27	23	20	17	12	20	17	14	9	14	11	8	9	5	3

The header of this table is structured as two rows spanning the overall label (α,β) or (β,α); the top sub-row values are 0.01, 0.01, 0.01, 0.01, 0.01, 0.025, 0.025, 0.025, 0.025, 0.05, 0.05, 0.05, 0.1, 0.1, 0.25 and the bottom sub-row values are 0.01, 0.025, 0.05, 0.1, 0.25, 0.025, 0.05, 0.1, 0.25, 0.05, 0.1, 0.25, 0.1, 0.25, 0.25.

Table A.2b Sample Sizes for the Wilcoxon Rank Sum Test
(Number of measurements to be performed in the reference area and in each survey unit)

	(α,β) or (β,α)													
	0.01	0.01	0.01	0.01	0.01	0.025	0.025	0.025	0.025	0.05	0.05	0.05	0.1	0.1	0.25
Δ/σ	0.01	0.025	0.05	0.1	0.25	0.025	0.05	0.1	0.25	0.05	0.1	0.25	0.1	0.25	0.25
0.10	5452	4627	3972	3278	2268	3870	3273	2646	1748	2726	2157	1355	1655	964	459
0.20	1370	1163	998	824	570	973	823	665	440	685	542	341	416	243	116
0.30	614	521	448	370	256	436	369	298	197	307	243	153	187	109	52
0.40	350	297	255	211	146	248	210	170	112	175	139	87	106	62	30
0.50	227	193	166	137	95	162	137	111	73	114	90	57	69	41	20
0.60	161	137	117	97	67	114	97	78	52	81	64	40	49	29	14
0.70	121	103	88	73	51	86	73	59	39	61	48	30	37	22	11
0.80	95	81	69	57	40	68	57	46	31	48	38	24	29	17	8
0.90	77	66	56	47	32	55	46	38	25	39	31	20	24	14	7
1.00	64	55	47	39	27	46	39	32	21	32	26	16	20	12	6
1.10	55	47	40	33	23	39	33	27	18	28	22	14	17	10	5
1.20	48	41	35	29	20	34	29	24	16	24	19	12	15	9	4
1.30	43	36	31	26	18	30	26	21	14	22	17	11	13	8	4
1.40	38	32	28	23	16	27	23	19	13	19	15	10	12	7	4
1.50	35	30	25	21	15	25	21	17	11	18	14	9	11	7	3
1.60	32	27	23	19	14	23	19	16	11	16	13	8	10	6	3
1.70	30	25	22	18	13	21	18	15	10	15	12	8	9	6	3
1.80	28	24	20	17	12	20	17	14	9	14	11	7	9	5	3
1.90	26	22	19	16	11	19	16	13	9	13	11	7	8	5	3
2.00	25	21	18	15	11	18	15	12	8	13	10	7	8	5	3
2.25	22	19	16	14	10	16	14	11	8	11	9	6	7	4	2
2.50	21	18	15	13	9	15	13	10	7	11	9	6	7	4	2
2.75	20	17	15	12	9	14	12	10	7	10	8	5	6	4	2
3.00	19	16	14	12	8	14	12	10	6	10	8	5	6	4	2
3.50	18	16	13	11	8	13	11	9	6	9	8	5	6	4	2
4.00	18	15	13	11	8	13	11	9	6	9	7	5	6	4	2

NUREG-1505

Table A.3 Critical Values for the Sign Test Statistic S+

N	0.005	0.01	0.025	0.05	0.1	0.2	0.3	0.4	0.5
4	4	4	4	4	3	3	3	2	2
5	5	5	5	4	4	3	3	3	2
6	6	6	5	5	5	4	4	3	3
7	7	6	6	6	5	5	4	4	3
8	7	7	7	6	6	5	5	4	4
9	8	8	7	7	6	6	5	5	4
10	9	9	8	8	7	6	6	5	5
11	10	9	9	8	8	7	6	6	5
12	10	10	9	9	8	7	7	6	6
13	11	11	10	9	9	8	7	7	6
14	12	11	11	10	9	9	8	7	7
15	12	12	11	11	10	9	9	8	7
16	13	13	12	11	11	10	9	9	8
17	14	13	12	12	11	10	10	9	8
18	14	14	13	12	12	11	10	10	9
19	15	14	14	13	12	11	11	10	9
20	16	15	14	14	13	12	11	11	10
21	16	16	15	14	13	12	12	11	10
22	17	16	16	15	14	13	12	12	11
23	18	17	16	15	15	14	13	12	11
24	18	18	17	16	15	14	13	13	12
25	19	18	17	17	16	15	14	13	12
26	19	19	18	17	16	15	14	14	13
27	20	19	19	18	17	16	15	14	13
28	21	20	19	18	17	16	15	15	14
29	21	21	20	19	18	17	16	15	14
30	22	21	20	19	19	17	16	16	15

Table A.3 Critical Values for the Sign Test Statistic S+ (continued)

N	α								
	0.005	0.01	0.025	0.05	0.1	0.2	0.3	0.4	0.5
31	23	22	21	20	19	18	17	16	15
32	23	23	22	21	20	18	17	17	16
33	24	23	22	21	20	19	18	17	16
34	24	24	23	22	21	19	19	18	17
35	25	24	23	22	21	20	19	18	17
36	26	25	24	23	22	21	20	19	18
37	26	26	24	23	22	21	20	19	18
38	27	26	25	24	23	22	21	20	19
39	27	27	26	25	23	22	21	20	19
40	28	27	26	25	24	23	22	21	20
41	29	28	27	26	25	23	22	21	20
42	29	28	27	26	25	24	23	22	21
43	30	29	28	27	26	24	23	22	21
44	30	30	28	27	26	25	24	23	22
45	31	30	29	28	27	25	24	23	22
46	32	31	30	29	27	26	25	24	23
47	32	31	30	29	28	26	25	24	23
48	33	32	31	30	28	27	26	25	24
49	33	33	31	30	29	27	26	25	24
50	34	33	32	31	30	28	27	26	25

For N larger than 50, the critical value, CV, may be obtained from the expression:

$CV = 0.5[N + z\sqrt{N}]$, where z is the $(1-\alpha)$ percentile of a standard normal distribution, which can be found in the following table:

α	0.005	0.01	0.025	0.05	0.1	0.2	0.3	0.4	0.5
z	2.575	2.326	1.96	1.645	1.282	0.842	0.524	0.253	0.000

Other values can be found in Table A.1.

Table A.4 Critical Values for the WRS Test

NOTE: When using this table under Scenario A, m is the number of reference area samples and n is the number of survey unit samples. When using this table for Scenario B, the roles of m and n in this table are reversed.

$m=2$, $n=$	2	3	4	5	6	7	8	9	10	11	12	13	14	15	16	17	18	19	20
$\alpha=0.001$	7	9	11	13	15	17	19	21	23	25	27	29	31	33	35	37	39	41	43
$\alpha=0.005$	7	9	11	13	15	17	19	21	23	25	27	29	31	33	35	37	39	40	42
$\alpha=0.01$	7	9	11	13	15	17	19	21	23	25	27	28	30	32	34	36	38	39	41
$\alpha=0.025$	7	9	11	13	15	17	18	20	22	23	25	27	29	31	33	34	36	38	40
$\alpha=0.05$	7	9	11	12	14	16	17	19	21	23	24	26	27	29	31	33	34	36	38
$\alpha=0.1$	7	8	10	11	13	15	16	18	19	21	22	24	26	27	29	30	32	33	35

$m=3$, $n=$	2	3	4	5	6	7	8	9	10	11	12	13	14	15	16	17	18	19	20
$\alpha=0.001$	12	15	18	21	24	27	30	33	36	39	42	45	48	51	54	56	59	62	65
$\alpha=0.005$	12	15	18	21	24	27	30	32	35	38	40	43	46	48	51	54	57	59	62
$\alpha=0.01$	12	15	18	21	24	26	29	31	34	37	39	42	45	47	50	52	55	58	60
$\alpha=0.025$	12	15	18	20	22	25	27	30	32	35	37	40	42	45	47	50	52	55	57
$\alpha=0.05$	12	14	17	19	21	24	26	28	31	33	36	38	40	43	45	47	50	52	54
$\alpha=0.1$	11	13	16	18	20	22	24	27	29	31	33	35	37	40	42	44	46	48	50

$m=4$, $n=$	2	3	4	5	6	7	8	9	10	11	12	13	14	15	16	17	18	19	20
$\alpha=0.001$	18	22	26	30	34	38	42	46	49	53	57	60	64	68	71	75	78	82	86
$\alpha=0.005$	18	22	26	30	33	37	40	44	47	51	54	58	61	64	68	71	75	78	81
$\alpha=0.01$	18	22	26	29	32	36	39	42	46	49	52	56	59	62	66	69	72	76	79
$\alpha=0.025$	18	22	25	28	31	34	37	41	44	47	50	53	56	59	62	66	69	72	75
$\alpha=0.05$	18	21	24	27	30	33	36	39	42	45	48	51	54	57	59	62	65	68	71
$\alpha=0.1$	17	20	22	25	28	31	34	36	39	42	45	48	50	53	56	59	61	64	67

$m=5$, $n=$	2	3	4	5	6	7	8	9	10	11	12	13	14	15	16	17	18	19	20
$\alpha=0.001$	25	30	35	40	45	50	54	58	63	67	72	76	81	85	89	94	98	102	107
$\alpha=0.005$	25	30	35	39	43	48	52	56	60	64	68	72	77	81	85	89	93	97	101
$\alpha=0.01$	25	30	34	38	42	46	50	54	58	62	66	70	74	78	82	86	90	94	98
$\alpha=0.025$	25	29	33	37	41	44	48	52	56	60	63	67	71	75	79	82	86	90	94
$\alpha=0.05$	24	28	32	35	39	43	46	50	53	57	61	64	68	71	75	79	82	86	89
$\alpha=0.1$	23	27	30	34	37	41	44	47	51	54	57	61	64	67	71	74	77	81	84

$m=6$, $n=$	2	3	4	5	6	7	8	9	10	11	12	13	14	15	16	17	18	19	20
$\alpha=0.001$	33	39	45	51	57	63	67	72	77	82	88	93	98	103	108	113	118	123	128
$\alpha=0.005$	33	39	44	49	54	59	64	69	74	79	83	88	93	98	103	107	112	117	122
$\alpha=0.01$	33	39	43	48	53	58	62	67	72	77	81	86	91	95	100	104	109	114	118
$\alpha=0.025$	33	37	42	47	51	56	60	64	69	73	78	82	87	91	95	100	104	109	113
$\alpha=0.05$	32	36	41	45	49	54	58	62	66	70	75	79	83	87	91	96	100	104	108
$\alpha=0.1$	31	35	39	43	47	51	55	59	63	67	71	75	79	83	87	91	94	98	102

$m=7$, $n=$	2	3	4	5	6	7	8	9	10	11	12	13	14	15	16	17	18	19	20
$\alpha=0.001$	42	49	56	63	69	75	81	87	92	98	104	110	116	122	128	133	139	145	151
$\alpha=0.005$	42	49	55	61	66	72	77	83	88	94	99	105	110	116	121	127	132	138	143
$\alpha=0.01$	42	48	54	59	65	70	76	81	86	92	97	102	108	113	118	123	129	134	139
$\alpha=0.025$	42	47	52	57	63	68	73	78	83	88	93	98	103	108	113	118	123	128	133
$\alpha=0.05$	41	46	51	56	61	65	70	75	80	85	90	94	99	104	109	113	118	123	128
$\alpha=0.1$	40	44	49	54	58	63	67	72	76	81	85	90	94	99	103	108	112	117	121

$m=8$, $n=$	2	3	4	5	6	7	8	9	10	11	12	13	14	15	16	17	18	19	20
$\alpha=0.001$	52	60	68	75	82	89	95	102	109	115	122	128	135	141	148	154	161	167	174
$\alpha=0.005$	52	60	66	73	79	85	92	98	104	110	116	122	129	135	141	147	153	159	165
$\alpha=0.01$	52	59	65	71	77	84	90	96	102	108	114	120	125	131	137	143	149	155	161
$\alpha=0.025$	51	57	63	69	75	81	86	92	98	104	109	115	121	126	132	137	143	149	154
$\alpha=0.05$	50	56	62	67	73	78	84	89	95	100	105	111	116	122	127	132	138	143	148
$\alpha=0.1$	49	54	60	65	70	75	80	85	91	96	101	106	111	116	121	126	131	136	141

Table A.4 Critical Values for the WRS Test (continued)

n =	2	3	4	5	6	7	8	9	10	11	12	13	14	15	16	17	18	19	20
m = 9																			
α=0.001	63	72	81	88	96	104	111	118	126	133	140	147	155	162	169	176	183	190	198
α=0.005	63	71	79	86	93	100	107	114	121	127	134	141	148	155	161	168	175	182	188
α=0.01	63	70	77	84	91	98	105	111	118	125	131	138	144	151	157	164	170	177	184
α=0.025	62	69	76	82	88	95	101	108	114	120	126	133	139	145	151	158	164	170	176
α=0.05	61	67	74	80	86	92	98	104	110	116	122	128	134	140	146	152	158	164	170
α=0.1	60	66	71	77	83	89	94	100	106	112	117	123	129	134	140	145	151	157	162

n =	2	3	4	5	6	7	8	9	10	11	12	13	14	15	16	17	18	19	20
m = 10																			
α=0.001	75	85	94	103	111	119	128	136	144	152	160	167	175	183	191	199	207	215	222
α=0.005	75	84	92	100	108	115	123	131	138	146	153	160	168	175	183	190	197	205	212
α=0.01	75	83	91	98	106	113	121	128	135	142	150	157	164	171	178	186	193	200	207
α=0.025	74	81	89	96	103	110	117	124	131	138	145	151	158	165	172	179	186	192	199
α=0.05	73	80	87	93	100	107	114	120	127	133	140	147	153	160	166	173	179	186	192
α=0.1	71	78	84	91	97	103	110	116	122	128	135	141	147	153	160	166	172	178	184

n =	2	3	4	5	6	7	8	9	10	11	12	13	14	15	16	17	18	19	20
m = 11																			
α=0.001	88	99	109	118	127	136	145	154	163	171	180	188	197	206	214	223	231	240	248
α=0.005	88	98	107	115	124	132	140	148	157	165	173	181	189	197	205	213	221	229	237
α=0.01	88	97	105	113	122	130	138	146	153	161	169	177	185	193	200	208	216	224	232
α=0.025	87	95	103	111	118	126	134	141	149	156	164	171	179	186	194	201	208	216	223
α=0.05	86	93	101	108	115	123	130	137	144	152	159	166	173	180	187	195	202	209	216
α=0.1	84	91	98	105	112	119	126	133	139	146	153	160	167	173	180	187	194	201	207

n =	2	3	4	5	6	7	8	9	10	11	12	13	14	15	16	17	18	19	20
m = 12																			
α=0.001	102	114	125	135	145	154	164	173	183	192	202	210	220	230	238	247	256	266	275
α=0.005	102	112	122	131	140	149	158	167	176	185	194	202	211	220	228	237	246	254	263
α=0.01	102	111	120	129	138	147	156	164	173	181	190	198	207	215	223	232	240	249	257
α=0.025	100	109	118	126	135	143	151	159	168	176	184	192	200	208	216	224	232	240	248
α=0.05	99	108	116	124	132	140	147	155	165	171	179	186	194	202	209	217	225	233	240
α=0.1	97	105	113	120	128	135	143	150	158	165	172	180	187	194	202	209	216	224	231

n =	2	3	4	5	6	7	8	9	10	11	12	13	14	15	16	17	18	19	20
m = 13																			
α=0.001	117	130	141	152	163	173	183	193	203	213	223	233	243	253	263	273	282	292	302
α=0.005	117	128	139	148	158	168	177	187	196	206	215	225	234	243	253	262	271	280	290
α=0.01	116	127	137	146	156	165	174	184	193	202	211	220	229	238	247	256	265	274	283
α=0.025	115	125	134	143	152	161	170	179	187	196	205	214	222	231	239	248	257	265	274
α=0.05	114	123	132	140	149	157	166	174	183	191	199	208	216	224	233	241	249	257	266
α=0.1	112	120	129	137	145	153	161	169	177	185	193	201	209	217	224	232	240	248	256

n =	2	3	4	5	6	7	8	9	10	11	12	13	14	15	16	17	18	19	20
m = 14																			
α=0.001	133	147	159	171	182	193	204	215	225	236	247	257	268	278	289	299	310	320	330
α=0.005	133	145	156	167	177	187	198	208	218	228	238	248	258	268	278	288	298	307	317
α=0.01	132	144	154	164	175	185	194	204	214	224	234	243	253	263	272	282	291	301	311
α=0.025	131	141	151	161	171	180	190	199	208	218	227	236	245	255	264	273	282	292	301
α=0.05	129	139	149	158	167	176	185	194	203	212	221	230	239	248	257	265	274	283	292
α=0.1	128	136	145	154	163	171	180	189	197	206	214	223	231	240	248	257	265	273	282

n =	2	3	4	5	6	7	8	9	10	11	12	13	14	15	16	17	18	19	20
m = 15																			
α=0.001	150	165	178	190	202	212	225	237	248	260	271	282	293	304	316	327	338	349	360
α=0.005	150	162	174	186	197	208	219	230	240	251	262	272	283	293	304	314	325	335	346
α=0.01	149	161	172	183	194	205	215	226	236	247	257	267	278	288	298	308	319	329	339
α=0.025	148	159	169	180	190	200	210	220	230	240	250	260	270	280	289	299	309	319	329
α=0.05	146	157	167	176	186	196	206	215	225	234	244	253	263	272	282	291	301	310	319
α=0.1	144	154	163	172	182	191	200	209	218	227	236	246	255	264	273	282	291	300	309

Table A.4 Critical Values for the WRS Test (continued)

m = 16	n =	2	3	4	5	6	7	8	9	10	11	12	13	14	15	16	17	18	19	20
	α=0.001	168	184	197	210	223	236	248	260	272	284	296	308	320	332	343	355	367	379	390
	α=0.005	168	181	194	206	218	229	241	252	264	275	286	298	309	320	331	342	353	365	376
	α=0.01	167	180	192	203	215	226	237	248	259	270	281	292	303	314	325	336	347	357	368
	α=0.025	166	177	188	200	210	221	232	242	253	264	274	284	295	305	316	326	337	347	357
	α=0.05	164	175	185	196	206	217	227	237	247	257	267	278	288	298	308	318	328	338	348
	α=0.1	162	172	182	192	202	211	221	231	241	250	260	269	279	289	298	308	317	327	336

m = 17	n =	2	3	4	5	6	7	8	9	10	11	12	13	14	15	16	17	18	19	20
	α=0.001	187	203	218	232	245	258	271	284	297	310	322	335	347	360	372	384	397	409	422
	α=0.005	187	201	214	227	239	252	264	276	288	300	312	324	336	347	359	371	383	394	406
	α=0.01	186	199	212	224	236	248	260	272	284	295	307	318	330	341	353	364	376	387	399
	α=0.025	184	197	209	220	232	243	254	266	277	288	299	310	321	332	343	354	365	376	387
	α=0.05	183	194	205	217	228	238	249	260	271	282	292	303	313	324	335	345	356	366	377
	α=0.1	180	191	202	212	223	233	243	253	264	274	284	294	305	315	325	335	345	355	365

m = 18	n =	2	3	4	5	6	7	8	9	10	11	12	13	14	15	16	17	18	19	20
	α=0.001	207	224	239	254	268	282	296	309	323	336	349	362	376	389	402	415	428	441	454
	α=0.005	207	222	236	249	262	275	288	301	313	326	339	351	364	376	388	401	413	425	438
	α=0.01	206	220	233	246	259	272	284	296	309	321	333	345	357	370	382	394	406	418	430
	α=0.025	204	217	230	242	254	266	278	290	302	313	325	337	348	360	372	383	395	406	418
	α=0.05	202	215	226	238	250	261	273	284	295	307	318	329	340	352	363	374	385	396	407
	α=0.1	200	211	222	233	244	255	266	277	288	299	309	320	331	342	352	363	374	384	395

m = 19	n =	2	3	4	5	6	7	8	9	10	11	12	13	14	15	16	17	18	19	20
	α=0.001	228	246	262	277	292	307	321	335	350	364	377	391	405	419	433	446	460	473	487
	α=0.005	227	243	258	272	286	300	313	327	340	353	366	379	392	405	419	431	444	457	470
	α=0.01	226	242	256	269	283	296	309	322	335	348	361	373	386	399	411	424	437	449	462
	α=0.025	225	239	252	265	278	290	303	315	327	340	352	364	377	389	401	413	425	437	450
	α=0.05	223	236	248	261	273	285	297	309	321	333	345	356	368	380	392	403	415	427	439
	α=0.1	220	232	244	256	267	279	290	302	313	325	336	347	358	370	381	392	403	415	426

m = 20	n =	2	3	4	5	6	7	8	9	10	11	12	13	14	15	16	17	18	19	20
	α=0.001	250	269	286	302	317	333	348	363	377	392	407	421	435	450	464	479	493	507	521
	α=0.005	249	266	281	296	311	325	339	353	367	381	395	409	422	436	450	463	477	490	504
	α=0.01	248	264	279	293	307	321	335	349	362	376	389	402	416	429	442	456	469	482	495
	α=0.025	247	261	275	289	302	315	329	341	354	367	380	393	406	419	431	444	457	470	482
	α=0.05	245	258	271	284	297	310	322	335	347	360	372	385	397	409	422	434	446	459	471
	α=0.1	242	254	267	279	291	303	315	327	339	351	363	375	387	399	410	422	434	446	458

Reject the null hypothesis if the test statistic is greater than the table (critical) value.

For n or m greater than 20, the table (critical) value can be calculated for Scenario A from

$$m(n+m+1)/2 + z\sqrt{nm(n+m+1)/12}$$

if there are few or no ties, and from

$$m(n+m+1)/2 + z\sqrt{\frac{nm}{12}\left[(n+m+1) - \sum_{j=1}^{g}\frac{t_j(t_j^2-1)}{(n+m)(n+m-1)}\right]}$$

if there are many ties, where g is the number of groups of tied measurements and t_j is the number of tied measurements in the jth group. For Scenario B, the roles of n and m are reversed in these equations. z is the $(1-\alpha)$ percentile of a standard normal distribution (see Table A.1).

Table A.5a   0.025 and 0.975 Percentiles of the Chi-Squared Distribution

DOF	0.025	0.975	DOF	0.025	0.975	DOF	0.025	0.975	DOF	0.025	0.975
			46	29.2	66.6	91	66.5	119.3	280	235.5	328.2
2	0.05	7.4	47	30	67.8	92	67.4	120.4	285	240.1	333.7
3	0.2	9.3	48	30.8	69.0	93	68.2	121.6	290	244.7	339.1
4	0.5	11.1	49	31.6	70.2	94	69.1	122.7	295	249.3	344.5
5	0.8	12.8	50	32.4	71.4	95	69.9	123.9	300	253.9	349.9
6	1.2	14.4	51	33.2	72.6	96	70.8	125.0	305	258.5	355.3
7	1.7	16.0	52	34.0	73.8	97	71.6	126.1	310	263.1	360.7
8	2.2	17.5	53	34.8	75.0	98	72.5	127.3	315	267.7	366.1
9	2.7	19.0	54	35.6	76.2	99	73.4	128.4	320	272.3	371.4
10	3.2	20.5	55	36.4	77.4	100	74.2	129.6	325	277.0	376.8
11	3.8	21.9	56	37.2	78.6	105	78.5	135.2	330	281.6	382.2
12	4.4	23.3	57	38.0	79.8	110	82.9	140.9	335	286.2	387.6
13	5.0	24.7	58	38.8	80.9	115	87.2	146.6	340	290.8	393.0
14	5.6	26.1	59	39.7	82.1	120	91.6	152.2	345	295.4	398.4
15	6.3	27.5	60	40.5	83.3	125	95.9	157.8	350	300.1	403.7
16	6.9	28.8	61	41.3	84.5	130	100.3	163.5	355	304.7	409.1
17	7.6	30.2	62	42.1	85.7	135	104.7	169.1	360	309.3	414.5
18	8.2	31.5	63	43.0	86.8	140	109.1	174.6	365	314.0	419.8
19	8.9	32.9	64	43.8	88.0	145	113.6	180.2	370	318.6	425.2
20	9.6	34.2	65	44.6	89.2	150	118.0	185.8	375	323.2	430.5
21	10.3	35.5	66	45.4	90.3	155	122.4	191.4	380	327.9	435.9
22	11.0	36.8	67	46.3	91.5	160	126.9	196.9	385	332.5	441.3
23	11.7	38.1	68	47.1	92.7	165	131.3	202.5	390	337.2	446.6
24	12.4	39.4	69	47.9	93.9	170	135.8	208.0	395	341.8	452.0
25	13.1	40.6	70	48.8	95.0	175	140.3	213.5	400	346.5	457.3
26	13.8	41.9	71	49.6	96.2	180	144.7	219.0	405	351.1	462.7
27	14.6	43.2	72	50.4	97.4	185	149.2	224.6	410	355.8	468.0
28	15.3	44.5	73	51.3	98.5	190	153.7	230.1	415	360.5	473.3
29	16.0	45.7	74	52.1	99.7	195	158.2	235.6	420	365.1	478.7
30	16.8	47.0	75	52.9	100.8	200	162.7	241.1	425	369.8	484.0
31	17.5	48.2	76	53.8	102.0	205	167.2	246.5	430	374.4	489.3
32	18.3	49.5	77	54.6	103.2	210	171.8	252.0	435	379.1	494.7
33	19.0	50.7	78	55.5	104.3	215	176.3	257.5	440	383.8	500.0
34	19.8	52.0	79	56.3	105.5	220	180.8	263.0	445	388.4	505.3
35	20.6	53.2	80	57.2	106.6	225	185.3	268.4	450	393.1	510.7
36	21.3	54.4	81	58.0	107.8	230	189.9	273.9	455	397.8	516.0
37	22.1	55.7	82	58.8	108.9	235	194.4	279.4	460	402.5	521.3
38	22.9	56.9	83	59.7	110.1	240	199.0	284.8	465	407.1	526.6
39	23.7	58.1	84	60.5	111.2	245	203.5	290.2	470	411.8	532.0
40	24.4	59.3	85	61.4	112.4	250	208.1	295.7	475	416.5	537.3
41	25.2	60.6	86	62.2	113.5	255	212.7	301.1	480	421.2	542.6
42	26.0	61.8	87	63.1	114.7	260	217.2	306.6	485	425.9	547.9
43	26.8	63.0	88	63.9	115.8	265	221.8	312.0	490	430.6	553.2
44	27.6	64.2	89	64.8	117.0	270	226.4	317.4	495	435.2	558.5
45	28.4	65.4	90	65.6	118.1	275	231.0	322.8	500	439.9	563.9

Table A.5b 0.05 and 0.95 Percentiles of the Chi-Squared Distribution

DOF	0.05	0.95	DOF	0.05	0.95	DOF	0.05	0.95	DOF	0.05	0.95
			46	31.4	62.8	91	70.0	114.3	280	242.2	320.0
2	0.1	6.0	47	32.3	64.0	92	70.9	115.4	285	246.9	325.4
3	0.4	7.8	48	33.1	65.2	93	71.8	116.5	290	251.6	330.7
4	0.7	9.5	49	33.9	66.3	94	72.6	117.6	295	256.2	336.1
5	1.1	11.1	50	34.8	67.5	95	73.5	118.8	300	260.9	341.4
6	1.6	12.6	51	35.6	68.7	96	74.4	119.9	305	265.5	346.7
7	2.2	14.1	52	36.4	69.8	97	75.3	121.0	310	270.2	352.1
8	2.7	15.5	53	37.3	71.0	98	76.2	122.1	315	274.9	357.4
9	3.3	16.9	54	38.1	72.2	99	77.0	123.2	320	279.6	362.7
10	3.9	18.3	55	39.0	73.3	100	77.9	124.3	325	284.2	368.0
11	4.6	19.7	56	39.8	74.5	105	82.4	129.9	330	288.9	373.4
12	5.2	21.0	57	40.6	75.6	110	86.8	135.5	335	293.6	378.7
13	5.9	22.4	58	41.5	76.8	115	91.2	141.0	340	298.3	384.0
14	6.6	23.7	59	42.3	77.9	120	95.7	146.6	345	303.0	389.3
15	7.3	25.0	60	43.2	79.1	125	100.2	152.1	350	307.6	394.6
16	8.0	26.3	61	44.0	80.2	130	104.7	157.6	355	312.3	399.9
17	8.7	27.6	62	44.9	81.4	135	109.2	163.1	360	317.0	405.2
18	9.4	28.9	63	45.7	82.5	140	113.7	168.6	365	321.7	410.5
19	10.1	30.1	64	46.6	83.7	145	118.2	174.1	370	326.4	415.9
20	10.9	31.4	65	47.4	84.8	150	122.7	179.6	375	331.1	421.2
21	11.6	32.7	66	48.3	86.0	155	127.2	185.1	380	335.8	426.5
22	12.3	33.9	67	49.2	87.1	160	131.8	190.5	385	340.5	431.8
23	13.1	35.2	68	50.0	88.3	165	136.3	196.0	390	345.2	437
24	13.8	36.4	69	50.9	89.4	170	140.8	201.4	395	349.9	442.3
25	14.6	37.7	70	51.7	90.5	175	145.4	206.9	400	354.6	447.6
26	15.4	38.9	71	52.6	91.7	180	150.0	212.3	405	359.4	452.9
27	16.2	40.1	72	53.5	92.8	185	154.5	217.7	410	364.1	458.2
28	16.9	41.3	73	54.3	93.9	190	159.1	223.2	415	368.8	463.5
29	17.7	42.6	74	55.2	95.1	195	163.7	228.6	420	373.5	468.8
30	18.5	43.8	75	56.1	96.2	200	168.3	234.0	425	378.2	474.1
31	19.3	45.0	76	56.9	97.4	205	172.9	239.4	430	382.9	479.3
32	20.1	46.2	77	57.8	98.5	210	177.5	244.8	435	387.6	484.6
33	20.9	47.4	78	58.7	99.6	215	182.1	250.2	440	392.4	489.9
34	21.7	48.6	79	59.5	100.7	220	186.7	255.6	445	397.1	495.2
35	22.5	49.8	80	60.4	101.9	225	191.3	261.0	450	401.8	500.5
36	23.3	51.0	81	61.3	103.0	230	195.9	266.4	455	406.5	505.7
37	24.1	52.2	82	62.1	104.1	235	200.5	271.8	460	411.3	511.0
38	24.9	53.4	83	63.0	105.3	240	205.1	277.1	465	416.0	516.3
39	25.7	54.6	84	63.9	106.4	245	209.8	282.5	470	420.7	521.5
40	26.5	55.8	85	64.7	107.5	250	214.4	287.9	475	425.5	526.8
41	27.3	56.9	86	65.6	108.6	255	219.0	293.2	480	430.2	532.1
42	28.1	58.1	87	66.5	109.8	260	223.7	298.6	485	434.9	537.3
43	29.0	59.3	88	67.4	110.9	265	228.3	304.0	490	439.7	542.6
44	29.8	60.5	89	68.2	112.0	270	232.9	309.3	495	444.4	547.9
45	30.6	61.7	90	69.1	113.1	275	237.6	314.7	500	449.1	553.1

Table A.6 1000 Random Numbers Uniformly Distributed Between Zero and One

0.382000	0.100681	0.596484	0.899106	0.884610	0.958464	0.014496	0.407422	0.863247	0.138585
0.245033	0.045473	0.032380	0.164129	0.219611	0.017090	0.285043	0.343089	0.553636	0.357372
0.371838	0.355602	0.910306	0.466018	0.426160	0.303903	0.975707	0.806665	0.991241	0.256264
0.951689	0.053438	0.705039	0.816523	0.972503	0.466323	0.300211	0.750206	0.351482	0.775658
0.074343	0.198431	0.064058	0.358348	0.487045	0.511216	0.373455	0.985900	0.040712	0.230720
0.004975	0.926145	0.100314	0.256691	0.775689	0.679647	0.809107	0.724326	0.085055	0.132267
0.756157	0.626514	0.173650	0.404798	0.552324	0.711509	0.555162	0.181158	0.970275	0.686941
0.528794	0.796686	0.805658	0.262215	0.177953	0.866756	0.114841	0.059511	0.761559	0.738395
0.986297	0.925596	0.903867	0.544969	0.500778	0.674978	0.489822	0.145787	0.037965	0.796258
0.671560	0.731681	0.584521	0.152226	0.892178	0.377819	0.200476	0.205786	0.333964	0.325144
0.300211	0.802179	0.696097	0.271493	0.904050	0.039125	0.709037	0.453719	0.516648	0.256539
0.291299	0.802149	0.789026	0.675954	0.755333	0.948515	0.619404	0.722068	0.968047	0.368603
0.850429	0.557054	0.873074	0.441053	0.217750	0.859035	0.280343	0.703299	0.707389	0.375835
0.329691	0.085971	0.976867	0.285531	0.534318	0.407392	0.997711	0.894711	0.810816	0.908597
0.574511	0.706076	0.401440	0.111026	0.897366	0.386334	0.095798	0.777642	0.783563	0.665731
0.656850	0.258461	0.765191	0.700308	0.858821	0.002808	0.678610	0.928831	0.042482	0.518143
0.912137	0.954314	0.594317	0.557665	0.968169	0.483016	0.255623	0.817896	0.496048	0.850642
0.668111	0.926939	0.451765	0.168096	0.061953	0.005158	0.541093	0.617603	0.492904	0.579455
0.601886	0.930052	0.533982	0.132054	0.082278	0.575915	0.829218	0.065676	0.270943	0.699698
0.414197	0.365581	0.435072	0.330088	0.211097	0.740471	0.523453	0.896786	0.603412	0.522782
0.589770	0.584918	0.497024	0.110508	0.593036	0.558916	0.774102	0.230811	0.731193	0.586718
0.545518	0.807337	0.964293	0.095370	0.108554	0.712271	0.883114	0.189917	0.015961	0.184545
0.580035	0.930620	0.162328	0.195105	0.677847	0.548387	0.294565	0.540819	0.173223	0.185308
0.852351	0.029842	0.250771	0.431745	0.544877	0.967467	0.724082	0.951018	0.570269	0.940641
0.259865	0.863244	0.883755	0.820887	0.041871	0.898312	0.420820	0.128208	0.030122	0.204718
0.682028	0.921291	0.472304	0.478530	0.584979	0.362407	0.823328	0.310556	0.990448	0.771569
0.729270	0.163274	0.808313	0.926084	0.232276	0.381664	0.090182	0.911985	0.852016	0.573138
0.014222	0.693411	0.700644	0.888028	0.168981	0.886013	0.287088	0.231574	0.042299	0.915372
0.167180	0.938932	0.121555	0.341258	0.095462	0.944060	0.511490	0.629475	0.835566	0.974364
0.475723	0.151585	0.284555	0.801660	0.808496	0.695242	0.068636	0.081332	0.442824	0.264687
0.265145	0.810297	0.200934	0.454268	0.408216	0.935545	0.093844	0.174780	0.433546	0.144322
0.075777	0.015259	0.719321	0.367351	0.660054	0.020234	0.878536	0.025666	0.302469	0.164434
0.459975	0.553606	0.958678	0.306223	0.213416	0.227607	0.721305	0.900418	0.075594	0.833796
0.943815	0.252419	0.533128	0.203742	0.756645	0.594531	0.518601	0.151555	0.382244	0.765648
0.496139	0.841884	0.155309	0.775964	0.892819	0.121097	0.654134	0.037446	0.531602	0.842860
0.849757	0.541856	0.223121	0.718528	0.678793	0.766930	0.171300	0.724631	0.946379	0.592608
0.350017	0.598224	0.965484	0.008179	0.506699	0.451033	0.838801	0.891781	0.949034	0.034455
0.789300	0.643147	0.771020	0.685781	0.446150	0.951476	0.675222	0.487564	0.491775	0.479080
0.046205	0.671194	0.576434	0.742393	0.432936	0.795495	0.906827	0.971435	0.095004	0.732383
0.414594	0.229041	0.770135	0.990143	0.911405	0.571306	0.318003	0.405896	0.136082	0.529832
0.397961	0.738731	0.101321	0.698172	0.835597	0.193823	0.644856	0.702261	0.835353	0.410321
0.592151	0.337565	0.695029	0.942595·	0.436842	0.153020	0.178442	0.767235	0.944792	0.263741
0.202582	0.846797	0.588214	0.673330	0.995331	0.281198	0.216346	0.887326	0.380352	0.078341
0.868526	0.098819	0.345164	0.075137	0.136021	0.727836	0.548448	0.040132	0.979186	0.594958
0.923246	0.489242	0.405713	0.511582	0.032014	0.245125	0.916623	0.566973	0.330088	0.272591
0.125889	0.894131	0.993255	0.820124	0.460524	0.195166	0.466231	0.399457	0.346446	0.808039
0.388287	0.634632	0.787317	0.595080	0.051607	0.510758	0.742241	0.435530	0.049532	0.055055
0.208197	0.602466	0.270211	0.980438	0.414136	0.806055	0.892422	0.323283	0.032533	0.183660

Table A.6 1000 Random Numbers Uniformly Distributed Between Zero and One (continued)

0.539445	0.606800	0.814722	0.019013	0.247505	0.560747	0.556291	0.470656	0.404859	0.929167
0.798975	0.091525	0.317698	0.397290	0.174932	0.421186	0.559557	0.168889	0.800104	0.430860
0.759972	0.225105	0.622150	0.103946	0.988800	0.796197	0.130375	0.932188	0.392621	0.077609
0.332774	0.608875	0.239357	0.721244	0.406964	0.701254	0.027192	0.133396	0.931364	0.582018
0.814356	0.110660	0.102451	0.402142	0.425642	0.845058	0.845119	0.385937	0.692495	0.829066
0.220954	0.229682	0.382153	0.693106	0.042573	0.884823	0.972777	0.505203	0.098636	0.507340
0.930357	0.565050	0.896664	0.854518	0.643971	0.223579	0.329630	0.909757	0.675314	0.759026
0.172857	0.017762	0.479110	0.840846	0.816828	0.310343	0.250435	0.612934	0.142796	0.704489
0.055086	0.927061	0.947508	0.254158	0.899716	0.758171	0.822321	0.136662	0.897244	0.332255
0.894345	0.055300	0.184576	0.998718	0.334727	0.338786	0.748619	0.377758	0.040986	0.519761
0.816065	0.814844	0.399274	0.672750	0.964843	0.672597	0.533036	0.293008	0.611682	0.492172
0.944273	0.998169	0.033418	0.554918	0.953581	0.832606	0.577563	0.875576	0.845332	0.457472
0.218421	0.158483	0.124149	0.005921	0.115513	0.747948	0.788903	0.886196	0.329905	0.103885
0.030519	0.312815	0.355022	0.834254	0.260231	0.847896	0.618427	0.815149	0.873440	0.733177
0.610187	0.430342	0.656911	0.294473	0.794671	0.460952	0.244270	0.199683	0.961852	0.656941
0.526933	0.851344	0.173956	0.521256	0.426038	0.695456	0.964629	0.338450	0.932218	0.727439
0.945738	0.978454	0.819025	0.976653	0.478957	0.506973	0.936644	0.250526	0.568072	0.133976
0.491592	0.920164	0.628040	0.531144	0.219459	0.027345	0.287332	0.049684	0.856014	0.496231
0.879940	0.870052	0.813654	0.728660	0.782586	0.614124	0.469192	0.366253	0.254555	0.452071
0.969054	0.499466	0.935179	0.234046	0.958464	0.105197	0.655873	0.678976	0.140599	0.994232
0.723197	0.150853	0.072329	0.049440	0.328562	0.054598	0.101932	0.036836	0.241157	0.381909
0.856441	0.769402	0.440474	0.438948	0.183264	0.906125	0.844386	0.286966	0.954039	0.167058
0.789361	0.926969	0.634968	0.245521	0.418592	0.759667	0.032044	0.198798	0.635029	0.352428
0.115635	0.598682	0.774468	0.043733	0.331706	0.222968	0.656362	0.417005	0.305551	0.569323
0.167089	0.174657	0.344340	0.972533	0.033326	0.444044	0.644856	0.606586	0.940641	0.045930
0.612262	0.152745	0.654134	0.757500	0.196844	0.415021	0.297617	0.100406	0.654347	0.663869
0.446913	0.685781	0.636311	0.989959	0.193823	0.646870	0.017029	0.448683	0.837245	0.190832
0.362743	0.277322	0.357433	0.801294	0.449965	0.965270	0.329722	0.121860	0.738639	0.459212
0.433882	0.053316	0.206793	0.284555	0.215369	0.094302	0.452803	0.863735	0.255135	0.457717
0.471084	0.813227	0.403394	0.993774	0.008576	0.500290	0.234413	0.450667	0.906888	0.640950
0.378277	0.104862	0.337260	0.219703	0.885647	0.291208	0.334605	0.513443	0.282022	0.865230
0.536180	0.499435	0.467940	0.127750	0.531877	0.429914	0.120792	0.244270	0.732139	0.853603
0.755211	0.689261	0.159276	0.865780	0.097385	0.747276	0.069582	0.868282	0.930692	0.601917
0.811060	0.620289	0.067843	0.044069	0.824061	0.861599	0.606433	0.330332	0.190893	0.289193
0.123814	0.223182	0.640248	0.168859	0.967864	0.744072	0.412915	0.362438	0.299387	0.670553
0.895444	0.052797	0.043214	0.308939	0.427168	0.946837	0.403211	0.573534	0.187719	0.238624
0.098300	0.903592	0.570605	0.113498	0.975494	0.049745	0.949370	0.074221	0.155797	0.260659
0.242744	0.870907	0.370037	0.901120	0.184881	0.315409	0.639912	0.033387	0.706595	0.762139
0.961303	0.992523	0.334574	0.535844	0.145756	0.486953	0.014252	0.087832	0.476363	0.255989
0.262673	0.162389	0.334208	0.814905	0.057863	0.753716	0.452040	0.319407	0.370617	0.135319
0.951018	0.080447	0.745293	0.743187	0.816828	0.760552	0.835383	0.193213	0.971679	0.783074
0.300851	0.215155	0.598010	0.985809	0.972686	0.783410	0.870754	0.049684	0.691305	0.226569
0.753624	0.734336	0.725364	0.596881	0.016511	0.451338	0.176397	0.080142	0.129551	0.206793
0.276589	0.728629	0.190497	0.444288	0.984863	0.296884	0.541551	0.819300	0.788263	0.462722
0.231910	0.436689	0.114872	0.213904	0.982208	0.582293	0.264016	0.682455	0.807215	0.479843
0.260292	0.207038	0.953032	0.104526	0.537889	0.905576	0.304727	0.584613	0.861049	0.538408
0.921445	0.221931	0.387249	0.944273	0.854060	0.662954	0.279092	0.829157	0.902921	0.251473
0.771294	0.190527	0.470717	0.094119	0.099124	0.157109	0.167852	0.870144	0.194250	0.526231
0.325571	0.916288	0.554521	0.779473	0.073763	0.618763	0.193701	0.858242	0.139470	0.642750
0.077822	0.480300	0.022187	0.031037	0.953795	0.278542	0.702658	0.703482	0.976897	0.543352
0.540727	0.387371	0.043519	0.849574	0.519364	0.511490	0.235298	0.349712	0.490188	0.093661
0.280221	0.014191	0.577441	0.154180	0.438429	0.600330	0.396466	0.685476	0.717460	0.630940

Table A.7a Values of r and k for the Quantile Test When α Is Approximately 0.01 [1]

m																				
	Number of Survey Unit Measurements, n																			
	5	10	15	20	25	30	35	40	45	50	55	60	65	70	75	80	85	90	95	100
5	r,k		11,11	13,13	16,16	19,19	22,22	25,25	28,28											r,k
	α		0.008	0.015	0.014	0.013	0.013	0.013	0.012											α
10		6,6	7,7	9,9	11,11	13,13	14,14	16,16	18,18	19,19	21,21	23,23	25,25	26,26	28,28	30,30				
		0.005	0.013	0.012	0.011	0.01	0.014	0.013	0.012	0.015	0.014	0.013	0.012	0.015	0.014	0.013				
15	3,3	7,6	6,6	7,7	8,8	10,10	11,11	12,12	13,13	15,15	16,16	17,17	18,18	19,19	21,21	22,22	23,23	24,24	26,26	27,27
	0.009	0.007	0.008	0.011	0.014	0.009	0.011	0.013	0.014	0.011	0.012	0.013	0.014	0.015	0.012	0.013	0.014	0.015	0.013	0.01
20	6,4	4,4	5,5	6,6	7,7	8,8	9,9	10,10	11,11	12,12	13,13	14,14	15,15	16,16	17,17	18,18	19,19	19,19	20,20	21,21
	0.005	0.008	0.009	0.01	0.011	0.011	0.011	0.011	0.011	0.011	0.011	0.012	0.012	0.012	0.012	0.012	0.012	0.015	0.015	0.01
25	4,3	7,5	4,4	5,5	6,6	7,7	8,8	9,9	9,9	10,10	11,11	12,12	12,12	13,13	14,14	15,15	16,16	16,16	17,17	18,18
	0.009	0.012	0.015	0.013	0.011	0.01	0.009	0.009	0.014	0.012	0.011	0.011	0.015	0.014	0.013	0.012	0.011	0.014	0.014	0.01
30	4,3	3,3	4,4	5,5	6,6	6,6	7,7	8,8	8,8	9,9	10,10	10,10	11,11	1211	12,12	13,13	14,14	14,14	15,15	15,15
	0.006	0.012	0.009	0.007	0.006	0.012	0.01	0.008	0.013	0.011	0.009	0.013	0.011	0.014	0.013	0.012	0.011	0.014	0.012	0.015
35	2,2	3,3	4,4	4,4	5,5	6,6	6,6	7,7	7,7	8,8	9,9	9,9	10,10	10,10	11,11	11,11	12,12	13,13	13,13	14,14
	0.013	0.008	0.006	0.014	0.01	0.007	0.012	0.009	0.014	0.011	0.009	0.013	0.01	0.014	0.011	0.015	0.012	0.011	0.013	0.012
40	2,2	3,3	7,5	4,4	5,5	5,5	6,6	6,6	7,7	7,7	8,8	8,8	9,9	9,9	10,10	10,10	11,11	11,11	12,12	12,12
	0.01	0.006	0.013	0.01	0.006	0.012	0.008	0.013	0.009	0.013	0.01	0.014	0.011	0.014	0.011	0.014	0.012	0.014	0.012	0.014
45	2,2	6,4	3,3	4,4	4,4	5,5	5,5	6,6	6,6	7,7	7,7	8,8	8,8	9,9	9,9	10,10	10,10	10,10	11,11	11,11
	0.008	0.008	0.013	0.007	0.014	0.008	0.014	0.009	0.013	0.009	0.013	0.009	0.012	0.009	0.012	0.009	0.012	0.015	0.012	0.01
50		4,3	3,3	4,4	4,4	5,5	5,5	5,5	6,6	6,6	7,7	7,7	8,8	8,8	8,8	9,9	9,9	10,10	10,10	10,10
		0.013	0.01	0.005	0.01	0.006	0.01	0.015	0.009	0.013	0.009	0.012	0.009	0.011	0.014	0.011	0.013	0.01	0.012	0.015
55		4,3	3,3	7,5	4,4	4,4	5,5	5,5	6,6	6,6	6,6	7,7	7,7	8,8	8,8	8,8	9,9	9,9	9,9	10,10
		0.01	0.008	0.013	0.008	0.014	0.007	0.011	0.007	0.01	0.014	0.009	0.012	0.008	0.01	0.013	0.009	0.012	0.014	0.011
60		4,3	3,3	3,3	4,4	4,4	5,5	5,5	5,5	6,6	6,6	6,6	7,7	7,7	7,7	8,8	8,8	8,8	9,9	9,9
		0.008	0.007	0.014	0.006	0.011	0.006	0.009	0.013	0.007	0.01	0.014	0.009	0.011	0.014	0.01	0.012	0.015	0.01	0.011
65		4,3	3,3	3,3	6,5	4,4	4,4	5,5	5,5	5,5	6,6	6,6	6,6	7,7	7,7	7,7	8,8	8,8	8,8	9,9
		0.007	0.006	0.012	0.006	0.009	0.013	0.007	0.01	0.014	0.008	0.011	0.014	0.009	0.011	0.014	0.009	0.011	0.014	0.011
70		2,2	6,4	3,3	7,5	4,4	4,4	5,5	5,5	5,5	5,5	6,6	6,6	6,6	7,7	7,7	7,7	8,8	8,8	8,8
		0.014	0.008	0.01	0.013	0.007	0.011	0.005	0.008	0.011	0.015	0.008	0.011	0.014	0.009	0.011	0.013	0.009	0.01	0.011
75		2,2	4,3	3,3	3,3	4,4	4,4	4,4	5,5	5,5	5,5	6,6	6,6	6,6	6,6	7,7	7,7	7,7	8,8.	8,8
		0.013	0.014	0.008	0.014	0.006	0.009	0.013	0.006	0.009	0.012	0.007	0.009	0.011	0.014	0.009	0.011	0.013	0.008	0.01
80		2,2	4,3	3,3	3,3	6,5	4,4	4,4	5,5	5,5	5,5	5,5	6,6	6,6	6,6	6,6	7,7	7,7	7,7	7,7
		0.011	0.012	0.007	0.012	0.006	0.008	0.011	0.005	0.007	0.01	0.013	0.007	0.009	0.012	0.014	0.009	0.01	0.013	0.01
85		2,2	4,3	3,3	3,3	7,5	4,4	4,4	4,4	5,5	5,5	5,5	5,5	6,6	6,6	6,6	6,6	7,7	7,7	7,7
		0.01	0.01	0.006	0.011	0.013	0.006	0.009	0.013	0.006	0.008	0.011	0.014	0.008	0.01	0.012	0.014	0.008	0.01	0.012
90			4,3	3,3	3,3	3,3	4,4	4,4	4,4	5,5	5,5	5,5	5,5	5,5	6,6	6,6	6,6	6,6	7,7	7,7
			0.009	0.005	0.009	0.014	0.005	0.008	0.011	0.005	0.007	0.009	0.012	0.015	0.008	0.01	0.012	0.014	0.008	0.019
95			4,3	6,4	3,3	3,3	6,5	4,4	4,4	4,4	5,5	5,5	5,5	5,5	6,6	6,6	6,6	6,6	6,6	7,7
			0.008	0.008	0.008	0.013	0.005	0.007	0.01	0.013	0.006	0.008	0.01	0.013	0.007	0.008	0.01	0.012	0.014	0.008
100	r,k		4,3	4,3	3,3	3,3	7,5	4,4	4,4	4,4	4,4	5,5	5,5	5,5	5,5	6,6	6,6	6,6	6,6	6,6
	α		0.007	0.014	0.007	0.011	0.013	0.006	0.008	0.011	0.015	0.007	0.009	0.011	0.013	0.007	0.008	0.01	0.012	0.01

[1] Values of the parameters r and k needed for the Quantile test calculated by Gilbert and Simpson (PNL-7409, 1992) for certain combinations of m (the number of measurements in the reference area) and n (the number of measurements in the survey unit). The value of α listed is that obtained from simulation studies.

Table A.7b Values of *r* and *k* for the Quantile Test When α Is Approximately 0.025

m		5	10	15	20	25	30	35	40	45	50	55	60	65	70	75	80	85	90	95	100
5	r,k	r,k		9,9	12,12	15,15	17,17	20,20	22,22	25,25											r,k
	α	α		0.03	0.024	0.021	0.026	0.024	0.028	0.025											α
10			7,6	6,6	8,8	9,9	11,11	12,12	14,14	17,17	18,18	20,20	21,21	23,23	24,24	26,26	27,27				
			0.029	0.028	0.022	0.029	0.024	0.029	0.025	0.025	0.029	0.026	0.029	0.026	0.029	0.026	0.029				
15		11,5	6,5	5,5	6,6	7,7	8,8	9,9	10,10	11,11	13,13	15,15	14,14	16,16	17,17	18,18	19,19	21,21	21,21	22,22	23,2
		0.03	0.023	0.021	0.024	0.026	0.027	0.028	0.029	0.03	0.022	0.023	0.023	0.024	0.025	0.025	0.026	0.021	0.027	0.027	0.02
20		8,4	3,3	4,4	5,5	6,6	7,7	12,11	13,12	9,9	10,10	11,11	12,12	13,13	13,13	14,14	15,15	16 16	17,17	17,17	18,1
		0.023	0.03	0.026	0.024	0.022	0.02	0.021	0.024	0.028	0.026	0.024	0.023	0.022	0.029	0.027	0.026	0.025	0.024	0.029	0.02
25		2,2	8,5	6,5	7,6	5,5	6,6	10,9	7,7	8,8	13,12	9,9	10,10	11,11	11,11	12,12	13,13	13,13	14,14	15,15	15,1
		0.023	0.027	0.021	6.023	0.025	0.02	0.026	0.027	0.023	0.027	0.027	0.024	0.022	0.028	0.025	0.823	0.628	0.025	0.023	0.02
30		6,3	6,4	9,6	4,4	7,6	5,5	9,8	6,6	7,7	12,11	8,8	9,9	9,9	10,10	10,10	11,11	11,11	12,12	13,13	13,1
		0.026	0.026	0.026	0.021	0.029	0.026	0.024	0.029	0.023	0.021	0.025	0.021	0.027	0.023	0.029	0.025	0.03	0.026	0.023	0.02
35		7,3	4,3	3,3	6,5	4,4	10,8	5,5	9,8	6,6	7,7	7,7	8,8	8,8	9,9	9,9	10,10	10,10	11,11	11,11	12,1
		0.03	0.03	0.023	0.02	0.026	0.022	0.027	0.024	0.027	0.02	0.027	0.021	0.027	0.022	0.027	0.022	0.027	0.022	0.027	0.02
40		3,2	4,3	8,5	11,7	6,5	4,4	10,8	5,5	9,8	6,6	10,9	7,7	12,11	8,8	8,8	9,9	9,9	10,10	10,10	11,1
		0.029	0.022	0.028	0.025	0.028	0.03	0.026	0.027	0.023	0.026	0.028	0.024	0.02	0.023	0.029	0.022	0.027	0.021	0.026	0.02
45		3,2	8,4	6,4	3,3	8,6	4,4	7,6	5,5	5,5	9,8	6,6	10,9	7,7	7,7	8,8	8,8	8,8	9,9	9,9	10,1
		0.023	0.029	0.036	0.026	0.021	0.023	0.025	0.02	0.028	0.023	0.024	0.026	0.022	0.027	0.02	0.025	0.03	0.023	0.027	0.02
50			2,2	6,4	3,3	11,7	6,5	4,4	7,6	5,5	5,5	9,8	6,6	6,6	7,7	7,7	12,11	8,8	8,8	13,12	9,9
			0.025	0.022	0.021	0.077	6.026	0.026	0.028	0.021	0.028	0.022	0.023	0.029	0.02	0.025	0.02	0.022	0.026	0.027	0.02
55			2,2	4,3	8,5	3,3	8,6	4,4	4,4	10,8	5,5	5,5	9,8	6,6	6,6	10,9	7,7	7,7	12,11	8,8	8,8
			0.022	0.029	0.028	0.028	0.021	0.02	0.029	0.021	0.022	0.028	0.022	0.092	0.028	0.029	0.023	0.027	0.023	0.023	0.02
60			14,5	4,3	8,5	3,3	11,7	6,5	4,4	7,6	10,8	5,5	5,5	9,8	6,6	6,6	10,9	7,7	7,7	7,7	8,8
			0.022	0.024	0.021	0.023	0.029	0.024	0.023	0.023	0.024	0.023	0.029	0.022	0.022	0.027	0.027	0.021	0.025	0.03	0.02
65			6,3	7,4	6,4	10,6	3,3	8,6	6,5	4,4	7,6	10,8	5,5	5,5	9,8	6,6	6,6	10,9	7,7	7,7	7,7
			0.028	0.021	0.025	0.025	0.029	0.021	0.029	0.026	0.026	0.026	0.023	0.029	0.022	0.021	0.026	0.026	0.020	0.024	0.02
70			6,3	2,2	6,4	8,5	3,3	13,8	6,5	4,4	4,4	7,6	10,8	5,5	5,5	9,8	6,6	6,6	6,6	10,9	7,7
			0.024	0.029	0.021	0.028	0.025	0.026	0.023	0.022	0.028	0.028	0.027	0.024	0.029	0.022	0.021	0.025	0.029	0.03	0.022
75			11,4	2,2	4,3	8,5	3,3	9,6	8,6	6,5	4,4	7,6	7,6	10,8	5,5	5,5	9,8	6,6	6,6	6,6	10,9
			0.022	0.026	0.028	0.022	0.022	0.028	0.021	0.027	0.024	0.023	0.03	0.029	0.024	0.029	0.021	0.021	0.024	0.028	0.02
80			7,3	2,2	4,3	6,4	10,6	3,3	13,8	6,5	4,4	4,4	7,6	10,8	5,5	5,5	5,5	9,8	6,6	6,6	6,6
			0.028	0.024	0.024	0.028	0.024	0.027	0.027	0.023	0.02	0.026	0.024	0.023	0.07	0.025	0.029	0.021	0.02	0.024	0.02
85			3,2	2,2	4,3	6,4	8,5	3,3	9,6	8,6	6,5	4,4	4,4	7,6	10,8	5,5	5,5	5,5	9,8	6,6	6,6
			0.029	0.021	0.021	0.023	0.028	0.023	0.03	0.02	0.026	0.022	0.028	0.026	0.024	0.021	0.025	0.029	0.02	0.02	0.02
90				5,3	11,5	9,5	8,5	3,3	3,3	13,8	6,5	6,5	4,4	4,4	7,6	10,8	5,5	5,5	5,5	9,8	9,8
				0.02	0.027	0.023	0.023	0.021	0.028	0.028	0.022	0.029	0.024	0.029	0.028	0.026	0.022	0.025	0.03	0.021	0.02
95				10,4	2,2	4,3	6,4	10,6	3,3	11,7	8,6	6,5	4,4	4,4	7,6	7,6	10,8	5,5	5,5	5,5	9,8
				0.029	0.029	0.028	0.029	0.023	0.025	0.026	0.02	0.025	0.021	0.026	0.024	0.029	0.027	0.022	0.026	0.03	0.02
100	r,k	r,k		6,3	2,2	4,3	6,4	8,5	3,3	3,3	13,8	6,5	6,5	4,4	4,4	7,6	10,8	10,8	5,5	5,5	5,5
	α	α		0.029	0.027	0.025	0.025	0.028	0.022	0.029	0.028	0.022	0.028	0.023	0.027	0.025	0.022	0.028	0.022	0.026	0.03

Table A.7c Values of *r* and *k* for the Quantile Test When α Is Approximately 0.05

m		5	10	15	20	25	30	35	40	45	50	55	60	65	70	75	80	85	90	95	100
								Number of Survey Unit Measurements, n												
5	r,k	r,k		8,8	10,10	13 13	15 15	17,17	19,19	21,21											r,k
	α	α		0.051	0.057	0.043	0.048	0.051	0.054	0.056											α
10	r,k		4,4	5,5	14,12	8,8	9,9	10,10	12,12	13,13	14,14	15,15	17,17	18,18	19,19	20,20	21,21	23,23			
	α		0.043	0.057	0.045	0.046	0.052	0.058	0.046	0.05	0.054	0.057	0.049	0.052	0.055	0.057	0.059	0.053			
15	r,k	2,2	3,3	4,4	5,5	6,6	7,7	8,8	9,9	9,9	10,10	11,11	12,12	13,13	14,14	15,15	16,16	16,16	17,17	18,18	19,1
	α	0.053	0.052	0.05	0.048	0.046	0.045	0.052	0.043	0.06	0.057	0.055	0.054	0.052	0.051	0.05	0.049	0.058	0.057	0.056	0.05
20	r,k	9,4	8,5	6,5	4,4	5,5	9,8	6,6	7,7	8,8	8,8	9,9	10,10	10,10	11,11	12,12	12,12	13,13	14,14	14,14	15,1
	α	0.04	0.056	0.04	0.053	0.043	0.052	0.056	48	0.043	0.057	0.051	0.046	0.057	0.052	0.048	0.057	0.053	0.049	0.057	0.05
25	r,k	6,3	6,4	3,3	6,5	4,4	5,5	5,5	6,6	11,10	7,7	8,8	8,8	9,9	9,9	10,10	11,11	11,11	11,11	12,12	12,1
	α	0.041	0.043	0.046	0.052	0.055	0.041	0.059	0.046	0.042	0.05	0.042	0.053	0.045	0.055	0.048	0.042	0.05	0.058	0.052	0.06
30	r,k	3,2	2,2	10,6	3,3	11,8	4,4	8,7	5,5	6,6	6,6	7,7	7,7	8,8	8,8	9,9	9,9	9,9	10,10	10,10	11,1
	α	0.047	0.058	0.052	0.058	0.045	0.056	0.044	0.054	0.04	0.053	0.041	0.052	0.042	0.051	0.042	0.05	0.059	0.049	0.057	0.04
35	r,k	8,3	2,2	6,4	3,3	6,5	4,4	4,4	8,7	5,5	9,8	6,6	6,6	7,7	7,7	8,8	8,8	8,8	9,9	9,9	10,10
	α	0.046	0.045	0.058	0.043	0.041	0.04	0.057	0.043	0.051	0.052	0.047	0.057	0.043	0.053	0.041	0.049	0.057	0.046	0.053	0.04
40	r,k	4,2	5,3	4,3	10,6	3,3	6,5	4,4	4,4	8,7	5,5	9,8	6,6	6,6	11,10	7,7	7,7	8,8	8,8	8,8	9,9
	α	0.055	0.048	0.057	0.059	0.053	0.048	0.043	0.058	0.042	0.048	0.047	0.042	0.051	0.042	0.045	0.053	0.041	0.048	0.055	0.04
45	r,k	4,2	9,4	2,2	8,5	3,3	8,6	6,5	4,4	4,4	8,7	5,5	5,5	9,8	6,6	6,6	11,10	7,7	7,7	8,8	8,8
	α	0.045	0.047	0.059	0.052	0.042	0.041	0.054	0.045	0.058	0.041	0.046	0.057	0.056	0.047	0.055	0.046	0.047	0.054	0.04	0.047
50	r,k		6,3	2,2	6,4	12,7	3,3	8,6	6,5	4,4	4,4	8,7	5,5	5,5	9,8	6,6	6,6	6,6	7,7	7,7	7,7
	α		0.051	0.05	0.051	0.05	0.049	0.049	0.059	0.047	0.059	0.041	0.045	0.054	0.051	0.043	0.05	0.058	0.041	0.048	0.05
55	r,k		3,2	2,2	4,3	8,5	3,3	5,4	6,5	9,7	4,4	4,4	8,7	5,5	5,5	9,8	6,6	6,6	6,6	11,10	7,7
	α		0.059	0.043	0.056	0.058	0.041	0.041	0.046	0.042	0.048	0.059	0.04	0.043	0.052	0.048	0.04	0.047	0.054	0.043	0.04
60	r,k		3,2	5,3	4,3	6,4	3,3	3,3	8,6	6,5	9,7	4,4	4,4	13,10	5,5	5,5	5,5	9,8	6,6	6,6	6,6
	α		0.052	0.052	0.046	0.059	0.035	0.047	0.043	51	0.046	0.049	0.059	0.052	0.042	0.05	0.058	0.054	0.044	0.05	0.055
65	r,k		.3,2	5,3	2,2	6,4	10,6	3,3	3,3	6,5	6,5	4,4	4,4	4,4	13,10	5,5	5,5	5,5	9,8	6,6	6,6
	α		0.045	0.043	0.053	0.048	0.05	0.04	0.052	0.041	0.055	0.042	0.05	0.06	0.052	0.041	0.048	0.055	0.051	0.041	0.04
70	r,k		8,3	9,4	2,2	4,3	8,5	5,4	3,3	3,3	6,5	6,5	4,4	4,4	4,4	13,10	5,5	5,5	5,5	9,8	9,8
	α		0.057	0.048	0.047	0.055	0.05	0.041	0.046	0.057	0.045	0.058	0.043	0.051	0.06	0.051	0.041	0.047	0.054	0.048	0.05
75	r,k		8,3	6,3	2,2	4,3	6,4	10,6	3,3	3,3	8,6	6,5	9,7	4,4	4,4	5,5	13,10	8,7	5,5	5,5	5,5
	α		0.049	0.056	0.043	0.047	0.054	0.053	0.04	0.051	0.044	0.049	0.041	0.044	0.052	0.06	0.051	0.047	0.046	0.052	0.05
80	r,k		4,2	6,3	5,3	2,2	6,4	8,5	5,4	3,3	3,3	6,5	6,5	9,7	4,4	4,4	7,6	13,10	8 7	5,5	5,5
	α		0.059	0.048	0.053	0.055	0.046	0.055	0.041	0.045	0.055	0.041	0.052	0.043	0.045	0.053	0.058	0.051	0.046	0.045	0.05
85	r,k		4,2	3,2	5,3	2,2	4,3	4,3	10,6	5,4	3,3	3,3	6,5	6,5	9,7	4,4	4,4	7,6	10,8	8,7	5,5
	α		0.054	0.058	0.047	0.05	0.054	0.048	0.056	0.049	0.049	0.059	0.044	0.055	0.046	0.046	0.053	0.059	0.06	0.045	0.04
90	r,k			3,2	5,3	2,2	6,4	6,4	8,5	5,4	3,3	3,3	8,6	6,5	6,5	4,4	4,4	4,4	7,6	10,8	8,7
	α			0.053	0.041	0.046	0.059	0.051	0.058	0.042	0.044	0.053	0.045	0.047	0.058	0.041	0.047	0.054	0.059	0.06	0.04
95	r,k			3,2	9,4	2,2	2,2	4,3	8,5	10,6	5,4	3,3	3,3	6,5	6,5	9,7	4,4	4,4	4,4	7,6	10,8
	α			0.048	0.048	0.042	0.056	0.059	0.05	0.058	0.048	0.048	0.056	0.041	0.05	0.040	0.042	0.048	0.054	0.59	0.059
100	r,k	r,k		3,2	6,3	5,3	2,2	4,3	6,4	10,6	5,4	3,3	3,3	3,3	6,5	6,5	9,7	4,4	4,4	4,4	7,6
	α	α		0.044	0.057	0.054	0.052	0.053	0.056	0.049	0.043	0.043	0.051	0.059	0.044	0.053	0.042	0.043	0.049	0.055	0.05

Table A.7d   Values of r and k for the Quantile Test When α Is Approximately 0.10

							Number of Survey Unit Measurements, n												
m	5	10	15	20	25	30	35	40	45	50	55	60	65	70	75	80	85	90	95	100
5	r,k		7,7	8,8	10,10	12,12	14,14	15,15	17,17											r,k
	α		0.083	0.116	0.109	0.104	0.1	0.117	0.112											α
10		3,3	4,4	5,5	6,6	7,7	8,8	9,9	10,10	11,11	12,12	13,13	14,14	15,15	16,16	1712	18,18			
		0.105	0.108	0.109	0.109	0.109	0.109	0.109	0.109	0.109	0.109	0.109	0.109	0.109	0.109	0.109	0.109			
15	9,4	10,6	3,3	4,4	5,5	5,5	6,6	7,7	7,7	8,8	9,9	9,9	10,10	11,11	11,11	12,12	13,13	13,13	14,14	15,1
	0.098	0.106	0.112	0.093	0.081	0.117	0.102	0.092	0.118	0.106	0.098	0.118	0.109	0.101	0.118	0.11	0.104	0.118	0.11	0.105
20	3,2	2,2	5,4	3,3	4,4	4,4	5,5	10,9	6,6	7,7	7,7	8,8	8,8	9,9	9,9	10,10	10,11	11,11	11,11	12,1
	0.091	0.103	0.093	0.115	0.085	0.119	0.093	0.084	0.099	0.083	0.102	0.088	0.105	0.092	0.107	0.095	0.108	0.098	0.11	0.1
25	4,2	7,4	8,5	3,3	3,3	4,4	4,4	8,7	5,5	10,9	6,6	6,6	7,7	7,7	8,8	8,8	8,8	9,9	9,9	10,1
	0.119	0.084	0.112	0.08	0.117	0.08	0.107	0.108	0.101	0.088	0.096	0.114	0.093	0.108	0.091	0.104	0.117	0.1	0.112	0.09
30	4,2.	5,3	2,2	14,8	3,3	3,3	9,7	4,4	8,7	5,5	5,5	6,6	6,6	6,6	7,7	7,7	7,7	8,8	8,8	8,8
	0.089	0.089	0.106	0.111	0.088	0.119	0.116	0.1	0.093	0.088	0.106	0.08	0.095	0.11	0.087	0.1	0.113	0.092	0.103	0.115
35	5,2	3,2	2,2	6,4	5,4	3,3	3,3	9,7	4,4	4,4	8,7	5,5	5,5	6,6	6,6	6,6	6,6	7,7	7,7	7,7
	0.109	0.119	0.086	0.12	0.091	0.093	0.12	0.112	0.094	0.114	0.107	0.094	0.11	0.081	0.094	0.107	0.12	0.094	0.105	0.11
40	5,2	3,2	5,3	2,2	12,7	5,4	3,3	6,5	9,7	4,4	4,4	8,7	5,5	5,5	5,5	6,6	6,6	6,6	6,6	7,7
	0.087	0.098	0.119	0.107	0.109	0.102	0.097	0.100	0.109	0.09	0.107	0.097	0.086	0.099	0.112	0.082	0.093	0.104	0.116	0.08
45	6,2	3,2	5,3	2,2	6,4	7,5	5,4	3,3	6,5	9,7	4,4	4,4	4,4	8,7	5,5	5,5	5,5	6,6	6,6	6,6
	0.103	0.082	0.094	0.091	0.115	0.086	0.112	0.1	0.101	0.107	0.087	0.102	0.117	0.107	0.091	0.103	0.115	0.083	0.093	0.103
50		7,3	9,4	7,4	2,2	10,6	5,4	3,3	3,3	6,5	9,7	4,4	4,4	4,4	8,7	5,5	5,5	5,5	5,5	6,6
		0.083	0.115	0.097	0.108	0.112	0.09	0.084	0.103	0.102	0.105	0.084	0.098	0.112	0.099	0.084	0.95	0.105	0.116	0.08
55		4,2	3,2	5,3	2,2	6,4	14,8	5,4	3,3	3,3	6,5	9,7	4,4	4,4	4,4	4,4	8,7	5,5	5,5	5,5
		0.109	0.114	0.114	0.095	0.112	0.111	0.098	0.088	0.104	0.103	0.104	0.082	0.095	0.107	0.12	0.107	0.088	0.098	0.103
60		4,2	3,2	5,3	2,2	2,2	8,5	5,4	5,4	3,3	3,3	6,5	9,7	4,4	4,4	4,4	4,4	8,7	5,5	5,5
		0.095	0.1	0.097	0.084	0.109	0.119	0.082	0.105	0.091	0.106	0.103	0.102	0.081	0.092	0.103	0.115	0.1	0.083	0.092
65		4,2	3,2	5,3	7,4	2,2	6,4	12,7	5,4	5,4	3,3	3,3	6,5	9,7	7,6	4,4	4,4	4,4	8,7	8,7
		0.084	0.089	0.082	0.090	0.097	0.11	0.113	0.089	0.111	0.093	0.108	0.104	0.101	0.084	0.09	0.1	0.11	0.094	0.107
70		5,2	7,3	9,4	5,3	2,2	2,2	8,5	7,5	5,4	3,3	3,3	3,3	6,5	9,7	7,6	4,4	4,4	4,4	4,4
		0.115	0.101	0.106	0.112	0.088	0.109	0.114	0.081	0.096	0.083	0.096	0.109	0.104	0.191	0.082	0.088	0.097	0.107	0.11
75		5,2	7,3	3,2	5,3	7,4	2,2	2,2	10,6	5,4	5,4	3,3	3,3	3,3	6,5	9,7	7,6	4,4	4,4	4,4
		103	0.088	0.111	0.098	0.101	0.099	0.119	0.117	0.083	0.102	0.085	0.098	0.11	0.105	0.1	0.081	0.086	0.095	0.10
80		5,2	4,2	3,2	5,3	7,4	2,2	2,2	8,5	14,8	5,4	5,4	3,3	3,3	3,3	6,5	6,5	9,7	4,4	4,4
		0.093	0.116	0.101	0.086	0.086	0.09!	0.109	0.111	0.11	0.089	0.107	0.088	0.099	0.111	0.105	0.12	0.116	0.084	0.09
85		5,2	4,2	3,2	9,4	5,3	2,2	2,2	2,2	10,6	7,5	5,4	5,4	3,3	3,3	3,3	6,5	6,5	9,7	4,4
		0.084	0.106	0.092	117	0.111	0.083	0.101	0.118	0.112	0.084	0.094	0.111	0.09	0.101	0.112	0.105	0.119	0.114	0.08
90			4,2	3,2	3,2	5,3	7,4	2,2	2,2	8,5	12,7	5,4	5,4	3,3	3,3	3,3	3,3	6,5	6,5	9,7
			0.097	0.085	0.119	0.099	0.095	0.093	0.109	0.108	0.114	0.083	0.099	0.082	0.092	0.102	0.113	0.105	0.119	0.11
95			4,2	7,3	3,2	5,3	7,4	2,2	2,2	2,2	10,6	14,8	5,4	5,4	3,3	3,3	3,3	3,3	6,5	6,5
			0.089	100	0.11	0.089	0.084	0.086	0.102	0.117	0.08	0.117	0.088	0.103	0.084	0.094	0.103	0.113	0.106	0.11
100	r,k		4,2	7,3	3,2	5,3	5,3	2,2	2,2	2,2	6,4	12,7	7,5	5,4	5,4	3,3	3,3	3,3	3,3	6,5
	α		0.082	0.09	0.102	0.08	0.109	0.08	0.095	0.11	0.118	0.109	0.086	0.093	0.08	0.086	0.095	0.104	0.114	0.10

NRC FORM 335
(2-89)
NRCM 1102,
3201, 3202

U.S. NUCLEAR REGULATORY COMMISSION

BIBLIOGRAPHIC DATA SHEET

(See instructions on the reverse)

1. REPORT NUMBER (Assigned by NRC, Add Vol., Supp., Rev., and Addendum Numbers, if any.)
NUREG-1505 Rev. 1

2. TITLE AND SUBTITLE

A Nonparametric Statistical Methodology for the Design and Analysis of Final Status Decommissioning Surveys

Interim Report for Use and Comment

3. DATE REPORT PUBLISHED	
MONTH	YEAR
June	1998

4. FIN OR GRANT NUMBER

5. AUTHOR(S)

C.V. Gogolak*, Environmental Measurements Laboratory
G.E. Powers, U.S. Nuclear Regulatory Commission
A.M. Huffert, U.S. Nuclear Regulatory Commission

6. TYPE OF REPORT

Technical

7. PERIOD COVERED *(Inclusive Dates)*

9/1/95 to 4/30/98

8. PERFORMING ORGANIZATION - NAME AND ADDRESS *(If NRC, provide Division, Office or Region, U.S. Nuclear Regulatory Commission, and mailing address; if contractor, provide name and mailing address.)*

*U.S. Department of Energy
Environmental Measurements Laboratory
201 Varick Street, 5th Floor
New York, NY 10014

Division of Regulatory Applications
Office of Nuclear Regulatory Research
U.S. Nuclear Regulatory Commission
Washington, DC 20555-0001

9. SPONSORING ORGANIZATION - NAME AND ADDRESS *(If NRC, type "Same as above"; if contractor, provide NRC Division, Office or Region, U.S. Nuclear Regulatory Commission, and mailing address.)*

Division of Regulatory Applications
Office of Nuclear Regulatory Research
U.S. Nuclear Regulatory Commission
Washington, DC 20555-0001

10. SUPPLEMENTARY NOTES

George E. Powers, NRC Project Manager

11. ABSTRACT *(200 words or less)*

This report describes a nonparametric statistical methodology for the design and analysis of final status decommissioning surveys in support of the final rulemaking on Radiological Criteria for License Termination published by the Nuclear Regulatory Commission in the Federal Register on July 21, 1997. The techniques described are expected to be applicable to a broad range of circumstances, but do not preclude the use of alternative methods as particular situations may warrant. Nonparametric statistical methods for testing compliance with decommissioning criteria are provided both for the case in which the radionuclides of concern occur in background and also for the case in which they do not occur in background. The tests described are the Sign test, the Wilcoxon Rank Sum test, and a Quantile test. These tests are performed in conjunction with an Elevated Measurement Comparison to provide confidence that the radiological criteria specified for license termination are met. The Data Quality Objectives process is used for the planning of final site surveys. This includes methods for determining the number of samples needed to obtain statistically valid comparisons with decommissioning criteria and the methods for conducting the statistical tests with the resulting sample data.

12. KEY WORDS/DESCRIPTORS *(List words or phrases that will assist researchers in locating the report.)*

Statistics, Decommissioning, Survey(s), Quality Assurance, Optimization, Nonparametric Statistics, Sign test, Wilcoxon Rank Sum test, Quantile test, Elevated Measurement Comparison

13. AVAILABILITY STATEMENT

unlimited

14. SECURITY CLASSIFICATION

(This Page)
unclassified

(This Report)
unclassified

15. NUMBER OF PAGES

16. PRICE

NRC FORM 335 (2-89)

This form was electronically produced by Elite Federal Forms, Inc.